Decolonizing Climate Adaptation

Ranjan Datta

Decolonizing Climate Adaptation

Indigenous Land-based Perspectives in Bangladesh

Springer

Ranjan Datta
Department of Humanities
Mount Royal University
Calgary, AB, Canada

ISBN 978-981-96-9139-5 ISBN 978-981-96-9140-1 (eBook)
https://doi.org/10.1007/978-981-96-9140-1

This Springer imprint is published by the registered company Springer Nature Singapore Pte Ltd.
The registered company address is: 152 Beach Road, #21-01/04 Gateway East, Singapore 189721, Singapore

This book is dedicated to the Indigenous and minority peoples of Bangladesh who are continuously struggling to protect their land rights, preserve their language, and sustain their traditional land-based culture and practices, advocating these as vital forms of climate action

Terminologies

Advocacy for land rights: Efforts to secure legal recognition and protection of Indigenous peoples' ancestral lands and rights against exploitation and displacement.

Agroforestry: A sustainable land-use practice combining agriculture and tree cultivation, used by the Marma for food security and ecological conservation.

Ancestral lands: Territories traditionally owned and occupied by Indigenous communities, holding cultural, spiritual, and ecological importance.

Anthropocene: A geological epoch characterized by significant human impact on Earth's ecosystems, including climate change and biodiversity loss.

Anthropogenic floods: Floods are intensified by human activities, such as deforestation and unregulated dam construction.

Biodiversity loss: The decline in the variety of life within ecosystems, often caused by human activities like monoculture plantations and habitat destruction.

Biodiversity refuges: Natural areas like sacred groves that preserve ecological diversity and provide climate resilience.

Climate adaptation strategies: Measures and practices adopted by communities to mitigate and cope with the impacts of climate change on their livelihoods and environment.

Climate justice: Advocacy for equitable climate policies that address systemic inequities and prioritize marginalized communities.

Climate justice frameworks: Approaches addressing the unequal burden of climate change on marginalized communities.

Climate resilience: The ability of communities, such as the Manipuri, to adapt to climate challenges through spiritual and ecological practices.

Colonial legacies: The enduring impacts of colonial practices, such as land dispossession and resource exploitation, on Indigenous communities.

Community resilience: The collective capacity of communities to rebuild and adapt during disasters, as seen in Marma-led cyclone center initiatives.

Community-based adaptation (CBA): Locally driven initiatives to strengthen resilience against environmental threats, tailored to specific regional contexts.

Community-based early warning systems: Locally driven initiatives to prepare for and respond to climate-induced disasters using traditional knowledge.

Community-based participatory action research (CBPAR): A collaborative framework empowering communities to identify local priorities and co-develop solutions.

Community-based resource management: Collective approaches to managing natural resources sustainably, ensuring equitable access and long-term resilience.

Community-driven strategies: Initiatives rooted in collective wisdom and collaboration within Indigenous communities to manage resources and mitigate environmental crises.

Community-led adaptation: Strategies developed and implemented by Indigenous communities to address climate change impacts, grounded in local knowledge and practices.

Cultural continuity: The preservation and transmission of Indigenous languages, practices, and traditions despite external pressures.

Cultural preservation: Efforts to maintain indigenous practices, languages, and identities amidst the challenges of modernization and displacement.

Cultural resilience: The ability of a community to sustain its cultural identity and practices in the face of external pressures or systemic challenges.

Cultural revitalization: Efforts to restore and sustain Indigenous cultural practices and identities in the face of environmental and systemic challenges.

Decolonial perspectives: Frameworks aimed at dismantling colonial structures to prioritize Indigenous voices and practices.

Decolonizing climate adaptation: An approach emphasizing the rectification of historical injustices and the integration of Indigenous perspectives into climate strategies.

Disaster preparedness: Proactive planning and resource management strategies aimed at reducing the impacts of natural disasters on vulnerable communities.

Discriminatory relief practices: Inequities in disaster aid distribution that marginalize Indigenous and minority communities.

Displacement: The forced or voluntary migration of people, often resulting in loss of ancestral lands, cultural heritage, and social cohesion.

Early marriage: A practice driven by poverty and cultural norms, often limiting educational and personal growth opportunities for girls.

Earth democracy: A concept advocating for ecological sustainability and social justice as foundational principles for development and governance.

Environmental awareness: The Marma communities' holistic understanding of ecological systems, shaping adaptive strategies for sustainable living.

Environmental conservation: Strategies like reforestation and sustainable watershed management to reduce flood risks.

Equitable resource distribution: Fair allocation of aid and resources to ensure marginalized groups receive adequate support during disasters.

Floating agriculture: A flood-resilient farming method using buoyant platforms to grow crops during high water levels.

Gender disparity: Inequalities in opportunities and rights between genders, particularly affecting education and leadership roles.

Gratitude practices: Daily rituals expressing appreciation for nature's resources, fostering sustainable living and community cohesion.

Human-created climate crisis: Environmental challenges exacerbated by human activities, such as deforestation, industrialization, and land encroachment.

Indigenous knowledge systems: Culturally rooted practices and wisdom that offer sustainable solutions to environmental challenges.

Indigenous land rights: The recognition and protection of Indigenous peoples' traditional territories and their rights to manage and sustain these lands.

Indigenous sovereignty: The right of Indigenous communities to self-govern and maintain control over their lands and resources.

Indigenous sovereignty: The right of Indigenous communities to self-determination and governance over their lands and resources.

Intergenerational knowledge transfer: The process of passing cultural, ecological, and spiritual wisdom to younger generations for continuity and resilience.

Intergenerational learning: Knowledge transfer between generations, fostering resilience and cultural continuity.

Land-based adaptation: Strategies grounded in traditional land use to enhance resilience against climate impacts.

Land-based knowledge: Knowledge systems rooted in the symbiotic relationship between Indigenous communities and their ancestral lands.

Land-based perspective: A worldview that recognizes the interconnectedness of living beings and the intrinsic value of natural, non-human-created resources.

Land-based spirituality: A worldview rooted in the interconnectedness of humans, nature, and the land, emphasizing harmony and reverence for natural elements.

Large-scale development projects: Infrastructure or industrial initiatives, often imposed without Indigenous consent, disrupting local ecosystems and communities.

Machang (elevated housing): Stilted homes designed by the Marma community to protect against rising floodwaters.

Mangrove afforestation: Planting mangrove trees to restore natural barriers, reduce cyclone impacts, and enhance coastal biodiversity.

Mangrove conservation: The protection and restoration of mangrove forests as natural buffers against cyclones and tidal surges.

Mangrove restoration: Replanting mangroves to buffer coastal areas from cyclones and tidal surges, enhancing ecosystem resilience.

Monoculture agriculture: An unsustainable farming practice focusing on a single crop, leading to soil degradation and biodiversity loss.

Nature-based solutions (NbS): Sustainable strategies like reforestation and wetland restoration to address ecological and social challenges effectively.

Panchashila: A land-based ethical code encompassing non-violence, truthfulness, and respect for life, applied to environmental conservation by the Rakhine community.

Participatory research: A collaborative research methodology that involves Indigenous communities in co-designing and co-authoring studies.

Relational theoretical framework: A research approach emphasizing interconnectedness, community priorities, and equitable partnerships between researchers and Indigenous participants.

Resilience: The capacity of communities, like the Marma, to adapt and recover from socio-economic and environmental adversities while maintaining cultural integrity.

Resource conservation practices: Sustainable methods, such as rainwater harvesting, for managing natural resources.

Rituals for agricultural sustainability: Ceremonies performed to honor land spirits, ensuring harmony with nature and promoting sustainable farming.

Sacred forests: Protected natural areas revered for their spiritual significance and ecological value, serving as reservoirs of biodiversity.

Sacred groves: Biodiversity-rich natural areas preserved for their spiritual significance and ecological importance.

Salinity intrusion: The process where saline water enters freshwater systems, disrupting agriculture and water supplies.

Seasonal dependency: Reliance on seasonal jobs or agriculture, leading to economic instability and social challenges.

Seasonal indicators: Environmental cues, such as plant flowering or animal behavior, used to guide agricultural practices and resource management.

Settler colonialism: The systemic replacement and marginalization of Indigenous communities through the establishment of non-Indigenous settlements and governance structures.

Spirituality: A guiding ethos in Marma culture emphasizing harmony with nature and stewardship of the environment through sacred practices and values.

Sustainable livelihoods: Economic activities aligned with environmental and cultural values to reduce dependency on exploitative industries.

Traditional ecological knowledge: Indigenous knowledge systems rooted in generations of interaction with the environment, critical for sustainability and disaster management.

Traditional housing techniques: Environmentally sustainable construction methods using renewable materials, rooted in Indigenous cultural practices.

Traditional land-based knowledge: Indigenous ecological and cultural knowledge passed down through generations, emphasizing harmony with nature.

Two-eyed seeing: Integrating Indigenous knowledge with Western science for holistic environmental solutions.

Umang Lai Haraoba: A Manipuri ritual honoring forest deities, symbolizing the interconnectedness of humans, nature, and spirituality.

Waterborne diseases: Illnesses such as diarrhea and cholera caused by contaminated floodwaters.

Youth self-determination: Empowering young people to lead their community's development and climate adaptation efforts.

Acknowledgments

As a decolonial and land-based minority scholar from Bangladesh and an anti-racist, racialized scholar from Canada, I extend my heartfelt gratitude to the Indigenous and minority peoples of Bangladesh whose relationships, guidance, and critical knowledge have critically shaped this book. Their continuous commitment to their land, culture, and community-driven climate justice efforts has been an inspiring force in this journey.

I am deeply grateful to the land-based Indigenous and minority communities across Bangladesh for their invaluable contributions. To the Indigenous Elders, knowledge-keepers, land-based educators, and youths, particularly from the Munda and Rakhain Indigenous communities in coastal areas, the Marma and Khyeng Indigenous communities in the Chittagong Hill Tracts, and the Monipuri Indigenous communities in the wetland regions, your wisdom, resilience, and generosity have provided the foundation for this research. Your teachings on land-based stewardship, cultural heritage, and climate justice have illuminated pathways toward sustainable and equitable futures.

Special thanks to the Indigenous research assistants who have played a vital role in this work. Barsha Kairy, from the Tea Garden in the wetland area, your dedication and tireless support have been instrumental in advancing this research. Your insights and connection to the local community have enriched the fieldwork and deepened my understanding of the challenges and opportunities faced by Indigenous peoples in these regions.

I am profoundly thankful to the Khyeng Indigenous land-based educators, including Hlaaung Kheyang, Hla Kray Prue Kheyang, and New Jai Kheyang, for their unwavering support and guidance. Your efforts in centering community knowledge and sustaining traditional practices have been both a source of inspiration and a critical resource for this project.

I am equally indebted to the leaders of the Munda Indigenous community, particularly Krishnapada Munda, the director of Sundarban Adibasi Munda Sangstha, for your continuous support. Your leadership and advocacy for the rights and recognition of the Munda people have provided a profound perspective on the intersection of community empowerment and climate justice.

Dr. Rajmoni Singha, your dedication and unwavering support throughout this research process have been invaluable. Your expertise and encouragement have greatly contributed to the depth and rigor of this work. I am equally grateful to Dr. Margot Hurlbert for accompanying me on some of the field journeys in the wetland areas. Your insights and collaboration have enriched the research experience and strengthened its outcomes.

Lastly, I express my sincere gratitude to the Canada Research Chair grants for funding this critical field research. Without this financial support, the depth and breadth of this work would not have been possible. Your commitment to advancing decolonial and climate justice research has been a cornerstone of this project.

This work stands as a testament to the collective knowledge, resilience, and advocacy of the Indigenous and minority communities of Bangladesh. It is through their enduring spirit and contributions that this research has taken shape. I am profoundly humbled and honored to have been part of this journey and am committed to continuing this collaborative path toward justice and sustainability.

Contents

About the Author

Ranjan Datta is serving as a Research Chair (CRC) in Community Disaster Research at Mount Royal University, Alberta, Canada, and has become a leading figure in research nationally and internationally. With over 18 years devoted to community-based research, Dr. Datta has made significant contributions across diverse domains, including community disaster research, community-led climate adaption, and environmental resource management, focusing with Indigenous and non-Indigenous communities in Canada, Ghana, Nepal, Bangladesh, and India. Dr. Datta's critical insights and impactful research derive from his extensive collaboration with diverse communities, showcasing his strong commitment to advancing knowledge and strengthening community-led disaster adaptions nationally and globally. His specialized knowledge in decolonial environmental sustainability and community-led governance positions him as a mentor to federally funded graduate and undergraduate students. Supported by a strong active network encompassing communities, scholars, and professionals, Dr. Datta's collaborative endeavors are primed for success. Internationally acclaimed, Dr. Datta has received prestigious awards such as the 2023 Sustainability Champion Award and the Excellence in Cultural Resource Management Award. Beyond academia, his engagement in grassroots movements reflects his dedication to social and environmental advocacy, exemplifying his commitment to transformative change. Through transdisiciplainary research and a steadfast commitment to global Indigenous sustainability, Dr. Datta continues to drive impactful research initiatives worldwide.

Chapter 1
Decolonizing Climate Adaptation: Indigenous Perspectives on Resilience and Justice

The growing impacts of the climate crisis disproportionately affect Indigenous communities worldwide, including those in Bangladesh. These communities, situated in environmentally vulnerable coastal, hilly, and wetland areas, bear the brunt of climate-induced disasters such as cyclones, floods, and landslides. Decolonizing climate adaptation strategies and integrating Indigenous land-based perspectives are critical for addressing these challenges. This approach rectifies historical injustices and leverages the traditional land-based knowledge and adaption practices embedded within Indigenous cultures and practices. This section explores the significance of decolonizing the climate crisis and emphasizes the importance of Indigenous perspectives in adaptation efforts. The book highlights the critical necessity of decolonizing climate adaptation from Indigenous traditional land-based culture and practice in Bangladesh. By addressing historical ongoing injustices, centering Indigenous traditional land-based knowledge and practices (TLBKP), and prioritizing community-led solutions, this book presents a critical case for reshaping the discourse on climate resilience. It advocates for recognizing and protecting Indigenous land rights as fundamental to creating sustainable climate adaptation strategies.

As the world faces the extremely negative impacts of many climate crises, this book offers a roadmap for integrating Indigenous wisdom into global adaptation efforts. It emphasizes the interconnectedness of Indigenous land rights, environmental justice, cultural heritage, and community resilience, highlighting the critical role of Indigenous knowledge in shaping a sustainable future. As climate vulnerabilities intensify, the lessons from Indigenous communities in Bangladesh resonate globally, offering transformative insights for building resilience and advancing environmental justice. Indigenous communities in Bangladesh face critical vulnerabilities due to the climate crisis and historical legacies of colonialism (Datta et al., 2024; Datta & Kairy, 2024). These include land dispossession, everyday oppression, extreme discrimination, cultural marginalization, and exclusion from decision-making processes. Studies have shown that Indigenous territories are

R. Datta, *Decolonizing Climate Adaptation*,
https://doi.org/10.1007/978-981-96-9140-1_1

disproportionately impacted by deforestation, soil degradation, and water scarcity—issues exacerbated by climate change (Datta, 2018; Garai et al., 2022). For instance, the Chittagong Hill Tracts, home to numerous Indigenous groups, experience frequent landslides due to deforestation driven by external economic interests (Datta et al., 2024; Datta & Kairy, 2024). Decolonizing climate adaptation requires addressing these structural inequities and advocating for Indigenous communities to lead adaptation efforts. This involves recognizing Indigenous land rights, implementing traditional knowledge into climate policy frameworks, and creating meaningful collaborations between Indigenous rightsholders and non-Indigenous stakeholders (Adnan, 2004; Datta et al., 2024; Datta & Kairy, 2024; Roy, 2000).

The TLBKP offers invaluable insights for sustainable climate adaptation. Rooted in centuries of land-based stewardship, this knowledge involves practices such as sustainable cultivation, water management, and disaster preparedness. For example, Indigenous communities in coastal Bangladesh use mangrove forests as natural buffers against cyclones and tidal surges, a practice supported by land-based studies (Haque et al., 2019). In the wetlands, Indigenous fishing techniques and water-sharing practices demonstrate how communities adapt to changing hydrological conditions. These practices not only enhance resilience but also promote biodiversity and land-based balance. Recognizing and integrating these methods into national adaptation strategies can enhance the effectiveness and cultural relevance of climate policies (Nath et al., 2016).

Indigenous communities in Bangladesh, particularly those residing in coastal and hilly regions, face acute climate challenges driven by geographical vulnerabilities, ecological changes, and socio-economic inequities. These challenges are further intensified by the enduring impacts of ongoing settler colonialism, including land dispossession, land grabbing, colonial farming practices, and external environmental governance imposed on Indigenous territories (Adnan, 2004; Chakma, 2016; Datta, 2020, 2022; Roy, 2000). Coastal and hilly areas have unique vulnerabilities due to the reliance of Indigenous communities on forest ecosystems and traditional agricultural practices, which are increasingly disrupted by deforestation and environmentally intrusive development projects. Forced deforestation, linked to outsiders' external economic interests, erodes natural buffers against landslides, destabilizes soil, and alters precipitation patterns, thereby undermining water availability and agricultural sustainability (Garai et al., 2022). These environmental transformations threaten traditional knowledge systems and practices, pushing Indigenous communities into a precarious position of balancing cultural preservation with climate adaptation (Chakma & Chakma, 2021).

Historical injustices with Indigenous and minority people in Bangladesh rooted in colonization also exacerbate these challenges. Encroachment on Indigenous lands, often driven by settler-colonial policies and infrastructural projects, disrupts traditional ways of life, displaces communities, and intensifies their exposure to the adverse effects of climate change (Adnan, 2004; Baird, 2008; Datta & Chapola, 2018; Roy, 2000). The complex and multilayer impacts of climate crises and historical dispossession necessitate a nuanced understanding of how colonial legacies shape contemporary climate vulnerabilities for Indigenous populations (Mason &

Rigg, 2019). Indigenous communities in Bangladesh maintain a significant connection to their ancestral lands, cultivating a deep, symbiotic relationship with their land over generations (Adnan, 2004; Datta, 2020; Roy, 2000). This connection incorporates knowledge of weather patterns, sustainable environmental practices, and environmental stewardship, all of which are critical for climate adaptation. However, the ongoing marginalization of Indigenous perspectives within mainstream climate policy underlines the need to decolonize climate discourse. This book critically examines these issues, advocating for Indigenous land rights and the incorporation of Indigenous knowledge systems into climate policies to foster sustainable and equitable adaptation strategies (Barkat et al., 2010; Haque et al., 2019; Jasanoff, 2018).

The book highlights the ethical and practical necessity of decolonizing climate adaptation in Bangladesh. By addressing historical injustices, centering Indigenous needs, and prioritizing community-based solutions, this book presents a compelling case for reshaping the discourse on climate resilience. It advocates for recognizing and protecting Indigenous land rights as fundamental to creating sustainable and meaningful adaptation strategies.

Decolonizing Climate Disasters

The historical and structural injustices faced by Indigenous communities in Bangladesh create a complex web of challenges, requiring targeted interventions grounded in a decolonial framework. Decolonizing climate disasters involves acknowledging and rectifying these historical injustices, dismantling systems of oppression, and ensuring land restitution for Indigenous communities (Chowdhury, 2008; Datta, 2018). This process emphasizes empowering Indigenous voices in decision-making and promoting meaningful actions in climate adaptation strategies (Chakma, 2016). Thus, this book sees Indigenous land rights as central to this decolonial approach. Since Indigenous land rights enable communities to continue sustainable land-use practices that preserve biodiversity and mitigate climate impacts. Recognizing land rights is not only a matter of justice but also a crucial foundation for implementing effective, culturally appropriate adaptation strategies rooted in Indigenous wisdom (Datta, 2020; Roy, 2000). By reframing climate disasters within a decolonial lens, this book advocates for a paradigm shift from top-down, technocentric governance to community-driven approaches that prioritize Indigenous knowledge and agency.

Relearning Adaptations

Relearning climate adaptation strategies from Indigenous TLBKP perspectives is essential for creating resilience and sustainability. Indigenous communities in Bangladesh have developed adaptive practices that draw on their critical understanding of local ecosystems, including sustainable traditional agricultural methods, community-based early warning systems, and traditional water management techniques (Datta, 2020; Nath et al., 2016). These practices not only enhance community resilience but also offer valuable insights for broader climate adaptation efforts. Thus, this book highlights the importance of integrating these Indigenous-led solutions into mainstream climate policies, bridging the gap between traditional knowledge and modern environmental governance. Creating collaboration between Indigenous TLBKP and non-Indigenous communities, policymakers, and researchers, focuses on the transformative potential of community-centric, culturally grounded approaches to climate resilience.

Challenging Colonial Legacies in Climate Policy

Ongoing colonial histories have critically shaped contemporary climate vulnerabilities for Indigenous communities in Bangladesh. Policies that prioritize economic development over environmental and cultural sustainability have resulted in the marginalization of Indigenous voices. For instance, large-scale infrastructural projects often encroach on Indigenous territories, displacing communities and disrupting traditional livelihoods (Datta, 2018; Roy, 2000).

Decolonizing climate adaptation necessitates dismantling these colonial legacies by centering Indigenous sovereignty and knowledge systems. This includes ensuring that Indigenous leaders have a voice in climate policy-making and that adaptation strategies are co-developed with affected communities (Battiste, 2013). Such an approach fosters meaningful and inclusivity, addressing the root causes of climate vulnerability rather than merely mitigating its symptoms.

Relational Approaches to Climate Resilience

Relational frameworks emphasize the interconnectedness of Indigenous people, land, and ecosystems. Indigenous perspectives inherently align with these frameworks, viewing land as a living being and emphasizing collective stewardship. This relational understanding is crucial for developing holistic adaptation strategies that account for both land-based and social dimensions of resilience (Datta, 2015; Smith, 2012; Wilson, 2016). In Bangladesh, relational approaches manifest in community-based early warning systems and collective resource management practices. For example, Indigenous women play a central role in maintaining TLBKP such as protecting seed banks and ensuring food security during climatic disruptions. These practices highlight the importance of TLBKP in building resilience (Garai et al., 2022).

Women in Indigenous communities are often disproportionately affected by climate change but also serve as key agents of adaptation. Their roles in agriculture, land-water management, and community leadership provide critical insights for addressing climate challenges (Datta et al., 2024). For instance, women-led initiatives in the Munda community have demonstrated how sustainable farming practices can enhance food security and mitigate climate risks (Datta et al., 2024; Datta & Kairy, 2024; Haque et al., 2019).

Thus, this book focuses on decolonizing climate adaptation from relational perspectives and involves recognizing and supporting these gendered dimensions of resilience. This includes ensuring that women's voices are heard in decision-making processes and that their contributions are valued in climate policies. Such an approach not only promotes gender rights but also strengthens the overall effectiveness of adaptation efforts (Battiste, 2013).

Building Cross-Cultural Collaborations

Decolonizing the climate crisis requires cross-cultural collaborations that meaningful integration between Indigenous TLBKP and non-Indigenous knowledge systems. These collaborations create respectful learning and innovation, creating hybrid solutions that are both scientifically strong and culturally appropriate. For example, implementing Indigenous practices of TLBKP with modern land-based research has shown promise in enhancing soil fertility and carbon sequestration (Nath et al., 2016). In Bangladesh, such collaborations can be facilitated through participatory research methodologies that prioritize Indigenous agencies. Co-designing projects and co-authoring publications with Indigenous community members ensure that their perspectives are authentically represented and respected (Datta, 2018).

Advocacy for Land Rights as a Climate Justice

Land rights are a keystone of climate justice for Indigenous communities. Secure land rights and TLBKP enable communities to implement sustainable practices, preserve biodiversity, and maintain cultural continuity. However, land grabbing and legal ambiguities often undermine these rights in Bangladesh, exacerbating climate vulnerabilities (Adnan, 2004). Advocating for land rights is integral to decolonizing climate adaptation. This includes legal recognition of Indigenous territories, restitution for historical injustices, and protection against future encroachments. Such measures not only address systemic inequities but also empower communities to lead adaptation efforts on their own terms (Roy, 2000). In this book, the lessons learned from decolonizing climate adaptation in Bangladesh have broader implications for global climate governance. Indigenous knowledge systems offer scalable solutions for sustainable development, biodiversity conservation, and disaster risk reduction. For instance, community-led mangrove restoration projects in Bangladesh provide a model for coastal resilience that can be applied in other vulnerable regions (Haque et al., 2019).

Moreover, centering Indigenous TLBKP perspectives challenges the dominant paradigms of climate governance, which often prioritize economic growth over environmental and social sustainability. By adopting decolonial approaches, policymakers can create more equitable and effective strategies for addressing the climate crisis (Battiste, 2013).

Thus, this book argues that decolonizing climate adaptation with Indigenous TLBKP is critical for addressing the multifaceted challenges of the climate crisis in Bangladesh. By recognizing the value of Indigenous knowledge, dismantling colonial legacies, and fostering equitable collaborations, researchers and policymakers can develop strategies that are culturally relevant, land-based sustainable, and socially just. This approach not only benefits Indigenous communities but also contributes to global efforts to build a more resilient and inclusive future.

Land-Based Theoretical Framework The land-based theoretical framework (i.e., TLBKP) emphasizes the intrinsic relationship Indigenous communities maintain with their ancestral lands. It is not merely utilitarian but deeply spiritual, cultural, and land based. For Indigenous peoples, land is their identity, encompassing generations of traditional knowledge and practices rooted in environmental stewardship (Datta, 2020). This framework is crucial for understanding the human-created climate crisis because it foregrounds the symbiotic relationship between humans and ecosystems, challenging exploitative practices that lead to environmental degradation.

In Bangladesh, Indigenous communities residing in coastal, hilly, and wetland areas exemplify land-based practices that offer sustainable solutions to climate adaptation. For instance, traditional water management systems, such as indigenous irrigation techniques and rainwater harvesting, are designed to work harmoniously with the natural water cycle. These practices have sustained agricultural productivity and ensured water availability in times of scarcity. Similarly, sustainable agriculture, which avoids monoculture and emphasizes crop rotation and biodiversity, enhances soil health and food security while reducing dependency on external inputs (Adnan, 2004; Barkat et al., 2010). This book highlights the importance of TLBKP frameworks, so researchers and policymakers take responsibility for how Indigenous TLBKP operate as adaptive mechanisms to the climate crisis. The TLBKP is not static but evolves in response to environmental changes, demonstrating resilience through practices like mangrove conservation and community reforestation. This book highlights the critical understanding and how integrating these practices into broader climate policies can help address the root causes of climate vulnerability while fostering equitable and sustainable solutions.

Decolonial Theoretical Framework The decolonial theoretical framework critiques the historical and ongoing impacts of colonialism on Indigenous communities, particularly their marginalization in environmental governance and land rights. This framework is essential for addressing human-created climate crises because it identifies the systemic inequities that exacerbate Indigenous vulnerabilities. Colonial legacies of land dispossession by the mainstream people particularly by mainstream Muslim people (Adnan, 2004), extractive industries, and imposed governance structures have disrupted Indigenous sovereignty and undermined their capacity to adapt to environmental changes (Datta, 2020). In Bangladesh, the effects of colonialism manifest through land grabbing, deforestation, and displacement driven by external economic interests and development projects (Adnan, 2004). These practices not only degrade ecosystems but also erode the cultural and spiritual connection Indigenous communities have with their land (Chowdhury, 2008; Datta, 2018). Decolonial approaches advocate for the restoration of land rights and the inclusion of Indigenous voices in decision-making processes as a means to rectify historical injustices and empower communities.

Through the decolonial framework, this book dismantles colonial legacies, the decolonial framework emphasizes the importance of land restitution and sovereignty. For example, Indigenous land rights in Bangladesh are directly linked to climate resilience, as secure tenure allows communities to implement sustainable land-use practices that mitigate climate impacts. This approach shifts the narrative from a technocratic focus on climate adaptation to one that prioritizes justice, and cultural relevance. The decolonial methodologies in this book encourage participatory research and collaboration, ensuring that Indigenous perspectives are not merely acknowledged but actively shape climate policies and practices.

Relational Theoretical Framework The relational theoretical framework emphasizes the interconnectedness of communities, ecosystems, and cultural practices in creating resilience. Unlike colonial approaches by the government and NGOs that isolate climate adaptation strategies, relational frameworks view adaptation as a collective endeavor that relies on the synergy between Indigenous TLBKP and non-Indigenous knowledge systems. This framework in this book is particularly important for understanding the human-created climate crisis because it highlights the value of relationships—between people, their environment, and their cultural heritage—in building resilience.

This book highlights how relational approaches are evident in community-based systems such as early warning networks for natural disasters. These networks rely on collective knowledge and action, integrating traditional land-based insights with modern technologies to enhance preparedness and response. For instance, Indigenous communities in coastal areas use their understanding of tidal patterns and weather indicators to predict and mitigate the impacts of cyclones and floods (Nath et al., 2016; Roy, 2000). Relational frameworks also emphasize the role of cultural practices in sustaining ecosystems. Traditional ceremonies, rituals, and spiritual practices that honor the land create a sense of collective responsibility and stewardship. These practices not only reinforce social cohesion but also enhance land-based sustainability by embedding conservation principles within cultural norms. Centering relational frameworks this book suggests that researchers and policymakers have the responsibility to design adaptation strategies that are both culturally grounded and land-based effective, bridging the gap between TLBKP and global environmental goals.

Indigenist Theoretical Framework In addition to all the theoretical frameworks, this book also highlights the Indigenist theoretical framework as a critical research framework for advocating the TLBKP epistemologies and worldviews, advocating for the reclamation and strengthening of their perspectives (Datta, 2022; Wilson, 2008). This framework challenges the marginalization of Indigenous knowledge systems within mainstream climate policies and emphasizes their relevance in addressing the human-created climate crisis. Indigenist approaches are critical because they prioritize self-determination, cultural integrity, and the decolonization of research methodologies. In Bangladesh, Indigenous leadership in sustainable agriculture illustrates the critical needs of Indigenous frameworks. TLBKP in Munda Indigenous communities, for example, play a pivotal role in designing and implementing adaptation strategies such as community gardens, seed preservation, and organic farming. These initiatives not only address food security but also empower the community as agents of change within their communities (Datta & Kairy, 2024). Indigenist frameworks also advocate for implementing TLBKP into science education and policy-making. For instance, Indigenous TLBKP about biodiversity, soil fertility, and water management can inform sustainable land-use practices and conservation efforts. By centering Indigenous perspectives, this framework promotes a holistic understanding of climate resilience that values cultural diversity and land-based balance. It also calls for ethical research practices that respect Indigenous sovereignty and foster reciprocal partnerships between researchers and communities.

Trans-Systematic Theoretical Framework The trans-systematic theoretical framework bridges disciplinary boundaries by integrating Indigenous, scientific, and cross-cultural perspectives to create holistic solutions (Datta, 2022; Smith, 2012). This framework is vital for addressing the human-created climate crisis because it transcends silos and fosters collaboration across knowledge systems. By recognizing the complementary strengths of TLBKP and modern approaches, trans-systematic frameworks promote inclusive and innovative adaptation strategies.

In Bangladesh, trans-systematic approaches can be seen in the blending of local agricultural techniques with modern environmental governance. For example, Indigenous practices such as crop diversification and agroforestry can be enhanced by scientific innovations like climate-resilient crop varieties and precision agriculture technologies (Garai et al., 2022; Haque et al., 2019). These hybrid solutions leverage the strengths of both knowledge systems, creating adaptable and context-specific strategies for climate resilience. Trans-systematic frameworks also emphasize cross-cultural dialogue and learning. By facilitating collaboration between Indigenous and non-Indigenous communities, these frameworks foster mutual respect and understanding, paving the way for shared solutions to global environmental challenges. Additionally, trans-systematic approaches advocate for interdisciplinary research that integrates insights from environmental science, sociology, anthropology, and Indigenous studies. This holistic perspective ensures that climate adaptation strategies are not only scientifically robust but also socially and culturally inclusive.

Thus, this book uses all these above theoretical frameworks for critically understanding how the human-created climate crisis poses significant challenges to Indigenous communities, yet their TLBKP offers valuable pathways for sustainable adaptation. All these theoretical frameworks discussed—land-based, decolonial, relational, Indigenist, and trans-systematic—provide critical lenses for understanding and addressing these challenges. Each framework underscores the importance of centering Indigenous perspectives, respecting their sovereignty, and implementing their knowledge systems into climate policies and practices. This book suggests By implementing these frameworks, researchers and policymakers have a responsibility to move beyond extractive and mainstream colonial approaches to climate adaptation, creating applicable solutions, culturally grounded, and environmentally sustainable. In the context of Bangladesh, where Indigenous communities face unique vulnerabilities, these frameworks highlight the transformative potential of Indigenous land-based learning and practices in building resilience and advancing environmental justice. Thus, this book TLBKP offers a roadmap for decolonizing climate adaptation and creating a more inclusive and sustainable future for all.

Research Methods in Decolonizing Climate Adaptation

In this book decolonial and TLBKP research methods play a pivotal role in understanding and addressing climate adaptation challenges through Indigenous land-based perspectives. This book explores five methodologies—deep listening, community-led land walks, traditional individual and collective story sharing, informal conversations, and drinking tea with informal conversations—and examines their application within the theoretical frameworks of land-based, decolonial, relational, Indigenist, and trans-systematic approaches. These methods align with TLBKP epistemologies and facilitate a critical understanding of climate adaptation in Bangladesh's coastal, hilly, and wetland areas.

Deep Listening Deep Listening involves an intentional and respectful engagement with Indigenous voices, allowing researchers to absorb knowledge conveyed through spoken words, silences, and embodied expressions (Datta et al., 2024). Rooted in relational and Indigenist theoretical frameworks, this method fosters reciprocal understanding and emphasizes the importance of hearing without imposing external interpretations (Wilson, 2016). In Bangladesh, Deep Listening is employed to capture the lived experiences of Indigenous communities confronting climate crises. For example, understanding how Indigenous women manage agricultural resilience or how spiritual beliefs shape adaptive practices requires researchers to listen deeply to personal narratives. By prioritizing Indigenous community needs, this method challenges colonial tendencies to marginalize local knowledge, aligning with decolonial frameworks that call for epistemologically meaningful (Datta, 2018). The practice also integrates TLBKP frameworks by respecting stories embedded in the physical and spiritual connection to land. For instance, the oral histories of communities in the Chittagong Hill Tracts (CHT) reveal how traditional water management practices, like bamboo irrigation, have sustained livelihoods and ecosystems for generations. Deep Listening enables researchers to contextualize these practices within broader climate adaptation strategies, promoting culturally grounded and land-based sustainable solutions (Haque et al., 2019).

Community-Led Land-Walks Community-led land walks involve guided tours of Indigenous land by the Elders and Knowledge-keepers, where community members share insights about their environment, traditional practices, and the impacts of climate change. This method resonates strongly with land-based and relational frameworks by emphasizing the connection between people and place and fostering collaborative knowledge-sharing.

In the wetlands of Bangladesh, land-walks led by Munda communities highlight the intricate relationships between mangroves, fisheries, and agricultural practices. These walks provide a platform for Indigenous experts to demonstrate adaptive techniques, such as the use of mangroves as natural buffers against storm surges. Through these interactions, researchers learn to appreciate the complex interplay between land-based resilience and cultural traditions (Garai et al., 2022). Community-led land-walks also align with decolonial frameworks by prioritizing Indigenous agencies and rejecting extractive research practices. Rather than imposing external frameworks, researchers learn directly from the land and its stewards, acknowledging Indigenous expertise as foundational to climate adaptation (Chowdhury, 2008). Additionally, Trans-Systematic approaches benefit from these walks by integrating traditional land-based knowledge with scientific methodologies, fostering innovative solutions to environmental challenges.

Traditional Individual and Collective Story Sharing TLBKP story sharing is a cornerstone of Indigenous knowledge transmission, serving as a medium for preserving histories, values, and land-based wisdom. Both individual and collective storytelling practices align with Indigenist and relational frameworks by emphasizing community cohesion and the relational nature of knowledge (Battiste, 2013; Wilson, 2008). In Bangladesh, storytelling sessions reveal the intergenerational strategies Indigenous communities use to adapt to environmental changes. For example, elders in coastal areas recount experiences of surviving cyclones and floods, emphasizing the importance of traditional warning systems and resource management techniques (Fig. 1.1). Such stories not only document resilience but also inspire younger generations to uphold sustainable practices (Nath et al., 2016). Story sharing supports Decolonial frameworks by challenging dominant narratives that often erase Indigenous perspectives. By foregrounding these voices, researchers contribute to a more inclusive understanding of climate adaptation. TLBKP stories rooted in land-based frameworks underline the spiritual and land-based

Fig. 1.1 Shows collective story sharing with Marma Indigenous community Elders and Knowledge-keepers at the Chittagong Hilly Tract (CHT) in Bangladesh in summer 2023. This photo was by the Indigenous research assistant Barsha Kairy

significance of Indigenous territories, reinforcing the need to protect these spaces as vital components of adaptation strategies (Roy, 2000).

Informal Conversations Informal conversations create a relaxed environment for dialogue, enabling participants to share insights without the constraints of formal research settings. This method is particularly effective within relational and Indigenist frameworks, as it fosters trust, reciprocity, and mutual learning (Wilson, 2016). In CHT regions of Bangladesh, informal conversations with Marma and Khyeng Indigenous communities provide valuable insights into how they navigate climate-induced challenges, such as landslides and deforestation. These discussions often uncover nuanced practices, like community reforestation initiatives, that might be overlooked in structured interviews. By engaging in informal dialogue, researchers build genuine relationships that enhance the depth and authenticity of their findings (Datta, 2022). Informal conversations also support trans-systematic approaches by facilitating interdisciplinary exchanges. For example, a conversation between Indigenous community members reveals opportunities to integrate traditional crop diversification techniques with modern agro-land-based innovations (Fig. 1.2). This meaningful collaborative spirit enriches both Indigenous and scientific knowledge systems, promoting holistic solutions to climate adaptation (Haque et al., 2019).

Fig. 1.2 Shows how drinking tea with Indiegnous community as effective research method for story sharing with Laitu Khyeng Indigenous community at the Chittagong Hilly Tract (CHT) in Bangladesh in summer 2023. This photo was taken by a Laitu Khyeng Indigenous youth

Drinking Tea The practice of sharing tea during conversations embodies relational and Land-Based frameworks by symbolizing hospitality, respect, and interconnectedness (Fig. 1.3). In Indigenous contexts, such gatherings are more than social rituals; they represent opportunities to exchange knowledge and build solidarity. In wetland areas of Bangladesh, tea-sharing sessions with Rakhine and Khyeng Indigenous communities often lead to reflective discussions about land-use practices and spiritual connections to the environment. For instance, conversations over tea might reveal how land-based spirituality informs adaptive measures like mangrove conservation. These moments of connection deepen researchers' understanding of the cultural dimensions of climate resilience (Garai et al., 2022). Drinking tea and engaging in informal conversations also align with decolonial and Indigenist frameworks by creating spaces where Indigenous voices are valued and respected. These interactions challenge hierarchical research models, fostering equal partnerships between researchers and community members. Additionally, they support Trans-Systematic methodologies by encouraging open, interdisciplinary dialogue that bridges cultural and epistemological divides (Wilson, 2016).

In this book, all these decolonial and TLBKP methodologies of deep listening, community-led land walks, traditional individual and collective story sharing, informal conversations, and drinking tea were integral to decolonizing climate

Fig. 1.3 Shows how drinking tea with Indiegnous community as effective research method for story sharing with Laitu Khyeng Indigenous community at the Chittagong Hilly Tract (CHT) in Bangladesh in summer 2023. This photo was taken by a Laitu Khyeng Indigenous youth

adaptation with Indigenous land-based perspectives. Each method aligns with theoretical frameworks that prioritize relationality, respect, and cultural integrity, offering a roadmap for inclusive and sustainable research practices. In Bangladesh, these methods provide invaluable insights into how Indigenous communities adapt to climate crises in coastal, hilly, and wetland areas. By centering Indigenous knowledge and fostering collaborative approaches, researchers and policymakers can develop strategies that respect cultural heritage, promote land-based sustainability, and address the root causes of climate vulnerability. Eventually, these methodologies contribute to a decolonized understanding of climate adaptation, emphasizing the transformative potential of Indigenous wisdom in shaping a resilient and equitable future.

Decolonial Research Ethics and Climate Adaptation

This book shows how ethical considerations are fundamental to research involving Indigenous communities, particularly when addressing sensitive issues like climate adaptation. This book explores key ethical practices—using Mount Royal University

(MRU) ethical guidelines, respecting cultural protocols and ceremonies, traditional gift sharing, honoring Indigenous knowledge, and participating in land rights and climate justice movements. These practices align with the theoretical frameworks of land-based, decolonial, relational, Indigenist, and trans-systematic approaches, ensuring culturally respectful and community-driven research in Bangladesh's coastal, hilly, and wetland areas.

Following Institutional Ethical Guidelines This book explains how Mount Royal University (MRU) emphasizes ethical research practices that prioritize respect, consent, and reciprocity. Following these guidelines is crucial in decolonizing climate adaptation research, as they ensure accountability and transparency. These principles align with the Indigenist and Decolonial frameworks by centering Indigenous sovereignty and rejecting extractive methodologies (Battiste, 2013; Wilson, 2016). MRU's ethical standards provide a framework for obtaining informed consent from Indigenous communities. Researchers are required to ensure that participants fully understand the purpose, methods, and implications of the study. This is particularly important in areas like the Chittagong Hill Tracts, where historical exploitation has fostered mistrust. By adhering to MRU guidelines, researchers build trust and foster equitable partnerships (Datta, 2018). Moreover, MRU's emphasis on minimizing harm aligns with the Relational framework by acknowledging the interconnectedness of people, land, and ecosystems. For example, ensuring that research activities do not disrupt local agricultural practices or spiritual ceremonies reflects a commitment to land-based and cultural sustainability.

Respecting Cultural Protocols and Traditional Ceremonies Respecting Indigenous cultural protocols and ceremonies is integral to ethical research. This book showcases how these practices are aligned with land-based and relational frameworks by recognizing the spiritual and cultural significance of Indigenous traditions. They also support Decolonial and Indigenist approaches by rejecting Eurocentric research paradigms and valuing Indigenous epistemologies (Chowdhury, 2008; Roy, 2000). This book suggests that researchers working with coastal, hilly, and wetland Indigenous communities must honor TLBKP cultural practices. For instance, before conducting research, researchers must take responsibility to center community needs, build trustful relationships. Researchers' responsibility is also useful to affirm the community's sovereignty and establish a reciprocal relationship between researchers and participants (Nath et al., 2016). Thus, this book explains how respecting cultural protocols also involves adapting research methods to align with community values. For example, in the Monda and Rakhine Indigenous communities, researchers might integrate spiritual practices into discussions about mangrove conservation. This approach not only enriches the research process but also strengthens the community's connection to their land and cultural heritage (Garai et al., 2022).

Traditional Gift Sharing Gift sharing is a longstanding tradition in many Indigenous cultures, symbolizing respect, gratitude, and reciprocity. This practice aligns with Relational and Indigenist frameworks by fostering mutual respect and strengthening relationships between researchers and communities (Wilson, 2016). In this book research, traditional gift sharing might involve offering locally meaningful items such as seeds, handmade crafts, or food. These gifts acknowledge the community's contributions to the research and express appreciation for their knowledge and hospitality. For instance, presenting traditional gifts to Elders in the CHT reinforces their role as knowledge keepers and demonstrates respect for their wisdom (Roy, 2000). Gift sharing also supported decolonial frameworks by challenging extractive research practices that take knowledge without giving back. It emphasized the importance of reciprocity, ensuring that research benefits are shared with the community. Additionally, this practice aligns with Land-Based frameworks by honoring the resources and traditions of Indigenous territories.

Honoring Indigenous Knowledge and Practices Respecting and centering Indigenous land-based knowledge and practices is a cornerstone of ethical research. This principle aligns with all five theoretical frameworks by emphasizing the value of Indigenous epistemologies and promoting equitable knowledge exchange (Battiste, 2013; Haque et al., 2019). In Bangladesh, honoring Indigenous knowledge involves integrating traditional land-based practices into climate adaptation strategies. For example, Indigenous communities' expertise in sustainable agriculture, water management, and disaster preparedness offers invaluable insights for addressing climate crises. Researchers must ensure that these contributions are recognized and incorporated into policy recommendations (Datta, 2018). This ethical practice also challenges colonial narratives that marginalize Indigenous knowledge. By centering on Indigenous perspectives, this book showcases how researchers contribute to a decolonized understanding of climate adaptation that respects cultural diversity and land-based integrity. For instance, the use of bamboo irrigation systems in the Chittagong Hill Tracts exemplifies how traditional practices can inform sustainable water management (Nath et al., 2016).

Being Active Parts in Land-Rights and Climate Justice Movements This book shows how ethical research extends beyond data collection to actively supporting Indigenous land rights and climate justice movements. This practice aligns with decolonial and Indigenist frameworks by addressing systemic inequities and advocating for Indigenous sovereignty (Chowdhury, 2008; Wilson, 2016). In this book research, the researcher shows how researchers can support land-rights movements by strengthening Indigenous voices in policy discussions and advocating for secure land tenure. For example, highlighting the impact of land grabbing on Munda communities in coastal areas underscores the need for legal protections and reparative justice (Roy, 2000). Similarly, supporting climate justice initiatives that prioritize Indigenous knowledge systems promotes equitable and sustainable adaptation strategies (Haque et al., 2019). Participation in these movements also reflects Relational and Trans-Systematic frameworks by fostering solidarity and collaboration across diverse stakeholders. By engaging with community-led initiatives, researchers contribute to a collective effort to address the root causes of climate vulnerability and promote resilience.

Thus, the decolonial ethical practices in this book research show how following MRU guidelines, respecting cultural protocols, engaging in traditional gift sharing, honoring Indigenous knowledge, and supporting land-rights movements are essential for decolonizing climate adaptation research. These practices align with theoretical frameworks that prioritize relationality, meaningful collaboration, and cultural integrity, ensuring that research benefits both Indigenous communities and the broader effort to address climate crises. In Bangladesh, these ethical considerations facilitate respectful and reciprocal engagement with Indigenous communities in coastal, hilly, and wetland areas. By centering Indigenous voices and fostering collaborative approaches, researchers and policymakers can develop strategies that promote land-based sustainability, cultural preservation, and social justice. Ultimately, these practices contribute to a decolonized understanding of climate adaptation, emphasizing the transformative potential of Indigenous knowledge in shaping a resilient and equitable future.

Researcher Positionality: Responsive and Transformative Research as Ceremony of Hopes

Engaging with Indigenous communities in research necessitates a transformative and responsive approach, challenging conventional Western notions of researchers' identity and purpose (Datta, 2018, 2022; Wilson, 2016). This book articulates how my research journey evolved into a transformative experience, shaping my identity as a decolonial transformational researcher. Decolonial transformational research became an act of self-determination for me, rooted in my experiences as a member of a minority family in Bangladesh. Growing up with everyday discrimination, oppression, and displacement due to minorities, researchers like myself often faced the burden of proving our legitimacy to both mainstream populations and scholars in Bangladesh. At research events, some scholars claimed exclusive rights to research in Bangladesh, going so far as to demand permission, under threat, for us to conduct research. Reclaiming our rights, including land and land-based cultural rights, often stirred opposition from mainstream scholars questioning the motives behind our identity affirmation and community advocacy.

Upon moving to Canada as a settler of colour immigrant scholar in Indigenous land, my expectations of a more positive and inclusive environment were met with new layers of discrimination from both Indigenous and non-Indigenous researchers. Despite receiving anti-racist, decolonial, and Indigenist research training, I encountered misconceptions and biases surrounding immigrant researchers' intersectional positions. Discrimination against racialized scholars, both Indigenous and non-Indigenous, is intricately connected to structural barriers rooted in resource scarcity, educational gaps, and limited opportunities.

Racism pervades not only across disciplines but within social science itself. As a community-based land-based researcher, I faced obstacles from my social science co-researchers, mirroring challenges encountered by mainstream researchers in Bangladesh. These barriers persist despite the need to prove ourselves to Indigenous and non-Indigenous scholars. Being a responsible researcher involves understanding these structural challenges and acknowledging the support, mentorship, and encouragement from anti-racist and decolonial scholars, community activists, and Indigenous community Elders. Individual efforts to change this system are commendable, but systematic change requires broader support for intersectional and anti-racist education within academia.

Responsive research, centred on community-led initiatives, has been healing and transformative for me. Establishing trustful relationships with local communities in Bangladesh and learning from land-based minority Elders and Knowledge-keepers has enriched my research process (Datta, 2018). In this collaborative approach, research becomes an action that creates hope, self-determination, and shared responsibility among researchers and communities.

Decolonial Research Analysis Process and Community Dissemination

Decolonial research necessitates collaborative, respectful, and culturally grounded methods for analyzing and disseminating findings. This book discusses how the key aspects of the decolonial research process—community knowledge sharing, thematic analysis, co-writing with Indigenous community members, and co-presenting—and examines their application within the theoretical frameworks of land-based, decolonial, relational, Indigenist, and trans-systematic approaches. This book explains how these practices are essential for ensuring that research on climate adaptation in Bangladesh's coastal, hilly, and wetland areas respects and amplifies Indigenous voices.

Community Knowledge Sharing Community knowledge sharing involves reciprocal exchanges of information between researchers and Indigenous communities, ensuring that findings are accessible and beneficial to the participants. Rooted in relational and land-based frameworks, this practice emphasizes the interconnectedness of people, ecosystems, and knowledge systems (Wilson, 2016). In Bangladesh, community knowledge sharing occurs through workshops, storytelling sessions, and participatory mapping. For example, we are working with Munda and Rakhine Indigenous communities in coastal regions may organize workshops to present findings on mangrove conservation while integrating community feedback. This iterative process validates Indigenous knowledge and ensures that research outcomes align with local priorities (Datta, 2018). This practice also supports decolonial frameworks by challenging traditional hierarchies in knowledge production. In this book, by centering Indigenous voices, researchers dismantle extractive methodologies and promote equitable partnerships. Community knowledge sharing in this book study aligned with Indigenist frameworks by honoring traditional knowledge as a vital component of climate adaptation strategies (Battiste, 2013).

Thematic Analysis In this book thematic analysis was used as a decolonial method by following Indigenous Elders, Knowledge-keepers, and land-based educators guidelines for identifying, analyzing, and interpreting patterns. When applied in decolonial research, this process must respect the cultural contexts and epistemologies of Indigenous participants. Thematic analysis aligns with Indigenist and trans-systematic frameworks by integrating Indigenous worldviews with academic methodologies (Chowdhury, 2008; Roy, 2000). In Bangladesh, thematic analysis is used to explore how Indigenous communities perceive and respond to climate challenges. For example, analyzing stories with Elders in the CHT may reveal recurring themes such as the spiritual significance of forests or the role of traditional agriculture in resilience. These themes inform culturally grounded adaptation strategies that reflect community priorities (Haque et al., 2019). The process also incorporates Relational frameworks by involving participants in data interpretation. Collaborative coding sessions, where researchers and community members jointly analyze transcripts, ensure that interpretations remain faithful to Indigenous perspectives. This approach fosters mutual learning and strengthens the validity of findings (Nath et al., 2016).

Co-Writing with Indigenous Community Members Co-writing was significant for Indigenous community members as active contributors to academic publications, ensuring their perspectives and knowledge are represented authentically. This practice aligned with decolonial and Indigenist frameworks by addressing power imbalances in knowledge production and amplifying Indigenous voices (Battiste, 2013; Wilson, 2016). In the context of Bangladesh, co-writing may involve Indigenous leaders or knowledge holders co-authoring articles on topics like sustainable agriculture or disaster resilience. For instance, a joint publication in different journals with all the community members highlighted the decolonial implications. Such collaborations ensure that Indigenous knowledge is recognized as a legitimate and valuable contribution to climate research (Datta, 2018). Co-writing supports land-based frameworks by embedding research findings within the cultural and land-based contexts of Indigenous territories. By integrating local narratives and traditional practices, these publications challenge dominant paradigms and promote alternative approaches to climate adaptation (Roy, 2000).

Co-Presenting Co-presenting involves researchers and Indigenous community members jointly delivering findings at conferences, workshops, and public forums. This practice exemplifies relational and trans-systematic frameworks by fostering cross-cultural dialogue and collaborative knowledge dissemination (Wilson, 2016). In this research, co-presenting allowed Indigenous representatives to share their experiences and insights directly with diverse audiences. For example, a co-presentation at an environmental conference might feature Indigenous leaders discussing traditional water management alongside researchers presenting scientific data. This collaborative approach not only enhances the credibility of Indigenous knowledge but also facilitates interdisciplinary solutions to climate challenges (Haque et al., 2019). Co-presenting also aligns with Decolonial frameworks by disrupting traditional power dynamics in academia. By sharing the stage with Indigenous partners, researchers acknowledge their equal role in knowledge production and dissemination. Furthermore, this practice reinforces Indigenist principles by centering Indigenous voices in discussions about their land and futures (Battiste, 2013).

Therefore, the decolonial research analysis process and community dissemination practices in this book research—community knowledge sharing, thematic analysis, co-writing with Indigenous community members, and co-presenting—are integral to decolonizing climate adaptation research. These methods align with theoretical frameworks that prioritize relationality, meaningfulness, and cultural integrity, ensuring that research outcomes benefit Indigenous communities and advance climate justice. In Bangladesh, these practices facilitate respectful and reciprocal engagement with Indigenous communities in coastal, hilly, and wetland areas. By centering Indigenous voices and fostering collaborative approaches, researchers and policymakers can develop strategies that promote land-based sustainability, and cultural preservation. Thus, these practices in this book contributed to a decolonized understanding of climate adaptation, emphasizing the transformative potential of Indigenous knowledge in shaping a resilient and equitable future. In a world facing climate change's complexities, this book offers a roadmap for integrating Indigenous wisdom into global adaptation efforts. It emphasizes the interconnectedness of environmental justice, cultural heritage, and community resilience, highlighting the critical role of Indigenous knowledge in shaping a more inclusive and sustainable future. As climate vulnerabilities intensify, the lessons from Indigenous communities in Bangladesh are applicable globally, offering transformative insights for building resilience and advancing environmental justice.

Chapter Descriptions

Chapter 2: *Settler Colonialism and Human-Created Disaster in Indigenous Communities in Bangladesh.* This chapter examines the compounded effects of settler colonialism and human-induced environmental crises on Indigenous communities in Bangladesh, focusing on the Chittagong Hill Tracts (CHT), coastal regions, and wetland areas. It delves into the multifaceted challenges these communities face, including land disputes, deforestation, large-scale development projects, and biodiversity loss—all intensified by inequities in policy-making and governance frameworks. These issues are rooted in historical and ongoing colonial practices that marginalize Indigenous populations, disrupt traditional land-based knowledge, and undermine sustainable practices. The chapter explores human-created factors such as land encroachment and industrial development that heighten climate vulnerabilities for Indigenous communities. It also centers the voices of Indigenous Elders and Knowledge-keepers, emphasizing traditional knowledge and cultural practices as vital strategies for climate adaptation and resilience. Finally, it advocates for transformative policy changes, including the recognition of Indigenous land rights, support for self-determination, and community-led approaches to climate governance. By addressing these objectives, the chapter underscores the need to decolonize climate governance and integrate Indigenous perspectives.

Chapter 3: *Human-Created Climate Crisis and the Significance of Land-Based Adaptation within and from Munda Indigenous Communities in Coastal Areas in Bangladesh.* This chapter investigates the implications of Indigenous knowledge systems in addressing human-created climate crises, emphasizing land-based practices, Indigenous sovereignty, and youth-led initiatives. It focuses on how the Munda Indigenous communities in coastal areas contend with challenges such as land grabbing, displacement, profit-oriented fish farming, deforestation, and the imposition of Western Forest management practices. The chapter highlights the role of traditional land-based knowledge and spirituality in fostering climate resilience and self-determination. Exploring intergenerational land-based learning, it underscores the significance of protecting cultural practices and promoting environmental stewardship across generations. Additionally, the chapter emphasizes Indigenous sovereignty in climate governance, positioning land rights as central to adaptation strategies. Through a decolonial lens, it critiques mainstream approaches to climate change and advocates for the integration of Indigenous knowledge into sustainable solutions, promoting both cultural revitalization and climate justice.

Chapter 4: *Munda Indigenous Youth Perspectives on Climate Crisis and Land-Based Adaptations.* This chapter illuminates the challenges and opportunities faced by Munda Indigenous youth amid the climate crisis. By centering their narratives, the chapter provides a platform for youth voices to advocate for culturally grounded, community-driven solutions. It explores the lived experiences, aspirations, and strategies of Munda youth, demonstrating how resilience and innovation emerge from adversity. Through their perspectives, the chapter contributes to a broader understanding of how Indigenous youth can lead the way in fostering sustainability and meaningfully in the face of climate challenges. The findings offer valuable lessons for global efforts to address the climate crisis while emphasizing the importance of intergenerational collaboration and Indigenous leadership.

Chapter 5: *Human-Created Flood and Land-Based Adaptation from Marma Indigenous Communities.* This chapter examines the severe impact of human-created floods on Marma Indigenous communities in the Chittagong Hill Tracts, highlighting the vulnerabilities and resilience strategies developed by these communities. Anthropogenic activities such as deforestation, unregulated dam construction, and climate change exacerbate flooding, disrupting livelihoods and cultural practices. The chapter explores the socioeconomic challenges faced by the Marma, including displacement, property loss, health crises, and discrimination in disaster relief. Despite these adversities, the Marma demonstrate resilience through land-based adaptations such as elevated housing, floating agriculture, and collective resource management. By preserving intergenerational Indigenous knowledge, they offer practical solutions to flood mitigation. The chapter advocates for integrating Marma strategies into climate policies, emphasizing meaningfully, environmental conservation, and resource transparency while showcasing their vital contributions to global adaptation efforts.

Chapter 6: *Resilience, Spirituality, and Environmental Awareness in the Face of Disasters: Perspectives from Marma Indigenous Communities*. This chapter explores the resilience, spirituality, and environmental awareness of Marma Indigenous communities in the Chittagong Hill Tracts as they navigate socio-environmental challenges and disasters. It highlights how their spiritual ethos, traditional land-based knowledge, and adaptive strategies foster harmony with nature and community solidarity. Key themes include the importance of sustainable practices, land-based perspectives, and cultural preservation in building resilience. The chapter underscores the role of inclusive governance and institutional accountability in addressing issues like displacement, corruption, and resource mismanagement. Integrating Indigenous knowledge into disaster risk reduction, offers actionable insights for promoting equitable and sustainable futures in vulnerable regions.

Chapter 7: *Land-Based Spirituality and Climate Adaptation in Khyeng Indigenous Communities.* This chapter examines the intersection of land-based spirituality and climate adaptation strategies among the Khyeng Indigenous communities in the Chittagong Hill Tracts. Anchored in cultural and land-based traditions, the Khyeng demonstrate resilience in the face of human-induced climate change. The chapter discusses challenges such as agricultural disruptions, health crises, and economic vulnerabilities while highlighting adaptive strategies rooted in spiritual practices. These include rituals for environmental harmony, seasonal agricultural preparations, and disaster preparedness initiatives. The findings advocate for integrating Indigenous knowledge into broader climate policies, showcasing how cultural and environmental dimensions converge to support sustainable adaptation strategies.

Chapter 8: *Land-Based Spirituality as Climate Resilience in Manipuri Indigenous Communities.* This chapter explores the role of land-based spirituality in fostering climate resilience within Manipuri Indigenous communities. It critiques mainstream environmental management approaches while emphasizing the transformative potential of Indigenous spiritual, cultural, and land-based practices. Themes such as sacred connections to land, climate change observations, and ritual-based environmental stewardship are examined in depth. The chapter highlights strategies for preserving Indigenous knowledge systems and cultural practices, advocating for their integration into global climate policies. By presenting actionable insights, it underscores the importance of Indigenous contributions to sustainability and climate justice, inspiring recognition of their value in addressing the climate crisis.

Chapter 9: *Challenging Human-Created Climate Crisis and Land-Based Spirituality as Climate Action from Coastal Area of Rakhine Indigenous Community Perspectives.* This chapter explores the human-created climate crisis from the perspective of the Rakhine Indigenous community, focusing on their land-based spirituality as a lens for resilience and climate action. Situated in coastal areas vulnerable to rising sea levels and extreme weather, the Rakhine community navigates challenges like deforestation, industrial waste, and land-based degradation. Themes include the anthropogenic causes of climate crises and the role of spiritual principles like Panchashila in fostering land-based stewardship. The chapter highlights practices such as afforestation, water conservation, and reverence for natural resources, demonstrating a holistic approach to resilience. It concludes by advocating for inclusive policies that integrate Indigenous voices and knowledge into climate adaptation strategies, balancing cultural heritage with adaptive practices.

Chapter 10: *Concluding Chapter*. This concluding chapter brings together the critical discussions of the book, highlighting the compounded effects of the climate crisis and settler colonialism on Indigenous communities in Bangladesh. It critiques Western, technocratic approaches to climate adaptation, which frequently marginalize Indigenous perspectives while prioritizing economic growth over land-based sustainability and cultural preservation. The chapter underscores the transformative potential of Indigenous land-based practices, rooted in land-based stewardship, spiritual connections, and intergenerational knowledge. Communities such as the Munda, Marma, Khyeng, and Rakhine exemplify these practices, offering sustainable and adaptive solutions in coastal, hilly, and wetland regions. A central focus of this chapter is the recognition of Indigenous land rights and self-determination as fundamental to achieving climate justice. It advocates for policies that integrate traditional land-based knowledge into national strategies, support community-led governance, and ensure culturally grounded education to empower future generations. Through the lens of decolonial learning, the chapter calls for dismantling colonial legacies embedded in climate governance and advancing inclusive frameworks that amplify Indigenous voices globally. By centering on Indigenous perspectives, the book emphasizes the urgency of fostering a sustainable and equitable future where Indigenous knowledge systems are pivotal in addressing the climate crisis and driving transformative change.

Chapter 2
Settler Colonialism and Human-Created Disaster in Indigenous Communities in Bangladesh

This chapter highlights the intensified impacts of settler colonialism and human-induced environmental crises on Indigenous communities in Bangladesh, with a focus on the Chittagong Hill Tracts (CHT), coastal regions, and wetland areas. It examines the multifaceted challenges these communities face, including land disputes, deforestation, large-scale development projects, and biodiversity loss, all of which are exacerbated by inequities in policy-making and governance frameworks. The chapter highlights how these issues are deeply rooted in historical and ongoing colonial practices that marginalize Indigenous populations, disrupt traditional land-based knowledge, and undermine sustainable practices. The chapter aims to achieve several key objectives. First, it explores the primary human-created factors, such as land encroachment, deforestation, and industrial development, that heighten climate vulnerabilities for Indigenous communities. Second, it centers the voices of Indigenous Elders and Knowledge-keepers to document their perspectives on these crises and the solutions they envision. Third, it emphasizes the importance of traditional land-based knowledge and cultural practices as critical strategies for climate adaptation and resilience. Finally, the chapter advocates for transformative policy changes, including the recognition of Indigenous land rights, support for self-determination, and the integration of community-led approaches into climate policies. By addressing these objectives, the chapter underscores the urgent need to decolonize climate governance and adopt collaborative approaches that amplify Indigenous voices and knowledge.

While Indigenous communities in Bangladesh have a rich history of environmental sustainability rooted in traditional land-based knowledge and practices, numerous challenges, primarily driven by mainstream human activities, have emerged. These challenges include land disputes, encroachment, uneven development, illegal settlements, deforestation, and biodiversity loss (Datta, 2020; Rahman & Alam, 2016; Seddon et al., 2020; Shaffril et al., 2020). Indigenous populations in the Chittagong Hill Tracts (CHT), coastal areas, and wetlands areas in Bangladesh face heightened risks associated with human-induced environmental crises, as

R. Datta, *Decolonizing Climate Adaptation*,
https://doi.org/10.1007/978-981-96-9140-1_2

indicated by various studies (Begum et al., 2019; Datta, 2020; Datta & Kairy, 2024; Mansoor et al., 2022; Shaffril et al., 2020). In their accounts, Indigenous communities attribute the impact of climate change to historical human interference, emphasizing that much of the crisis could have been avoided or controlled without such interference.

This chapter showcases systemic ongoing challenges confronting Indigenous communities in accessing decision-making processes and resources to adapt and mitigate the effects of climate change policies and practices. Inequitable representation, limited participation, and power imbalances within policy and governance frameworks pose obstacles to effective responses to climate change impacts, limiting the integration of Indigenous knowledge and perspectives. Addressing these challenges necessitates the recognition and respect of Indigenous rights, support for self-determination, and the adoption of collaborative and inclusive approaches to climate change mitigation and adaptation. The study, guided by a relational theoretical research framework (Datta, 2020), created the origins of the climate change crisis in Indigenous communities in the Chittagong Hill Tracts, highlighting the role of human interference. It explores the impacts of human-induced climate change on these communities and seeks to understand how Indigenous groups envision community-led and community-engaged solutions integrated into their daily practices.

Indigenous Community Perspectives on Human-Created Climate Crisis and Community-led Solutions

This chapter focuses on the CHT, coastal areas, and wetlands areas in Bangladesh, aiming to address the heightened climate change threats Indigenous communities face, especially those belonging to smaller Indigenous communities. Various human-created challenges have significantly impacted traditional environmental sustainability among Indigenous communities. This study has shown how specific human-created reasons contribute to climate change in these communities. For instance, this study shared some of the significant reasons from the communities regarding communities thinks why climate change is human created in their communities (refer to Fig. 2.1).

Land Disputes and Encroachment Indigenous community Elders and Knowledge-keepers emphasize that many of the current climate challenges faced by their communities stem from severe land encroachments and settlements by non-Indigenous individuals. This has resulted in the displacement of Indigenous communities from their ancestral lands. Beyond illegal settlements, various forms of land grabbing, such as large-scale infrastructure projects, commercial agriculture, and illegal logging, contribute significantly to the encroachment. An Indigenous Elder highlighted, “Many small Indigenous communities are profoundly affected by these ongoing land disputes initiated by outsiders.”

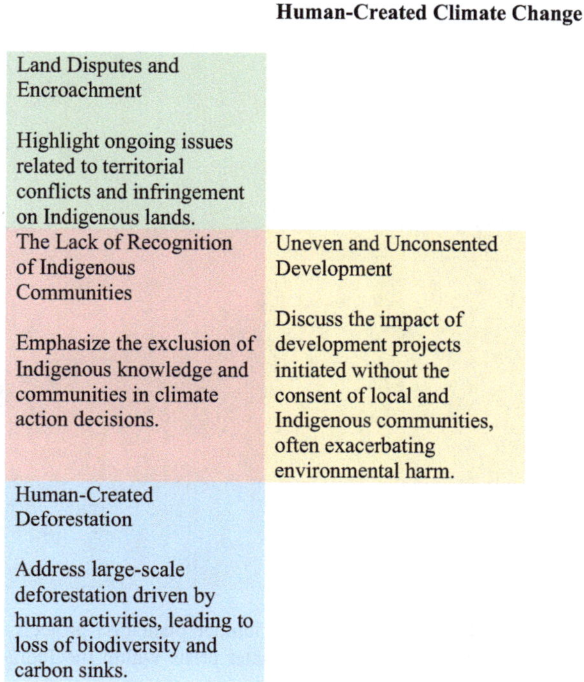

Fig. 2.1 Shows how many climate change crisis have been human-created in the Indigenous communities at the CHT, Bangladesh

Lack of Indigenous Land-Right Recognition Indigenous Elders pointed out that the absence of recognition of Indigenous land rights has allowed non-Indigenous individuals to occupy their lands and diminish centuries-old Indigenous traditional land-based cultivation knowledge and practices. Another Knowledge-keeper noted that the absence of customary land rights, coupled with limited participation in policy discussions, intensifies their vulnerability and restricts their ability to safeguard their environment and culture. Indigenous communities often experience marginalization and exclusion from decision-making processes related to land and natural resource management.

Human-Created Deforestation Human-induced deforestation stands out as a primary challenge contributing to climate change in many Indigenous communities in Bangladesh. Elders explained that much of the deforestation is driven forcefully by non-Indigenous individuals and industries (refer to Fig. 1.2). A knowledge keeper detailed, "Human-created deforestation takes various forms, primarily driven by illegal logging, shifting cultivation, and the establishment of commercial tobacco, break fields, and plantations in the hilly areas (Fig. 2.2). All these human-induced activities, primarily by outsiders, exacerbate the loss of traditional forests, severely impacting Indigenous environmentally sustainable livelihoods, which rely on forest resources for traditional practices, agriculture, non-timber forest product gathering, language, and cultural activities." Another Elder highlighted how human-induced deforestation significantly amplifies climate change, leading to soil erosion, a severe food crisis, and the loss of wildlife habitats.

Fig. 2.2 Illustrates how external lumber plantation projects have encroached upon traditional land-based forests, triggering a water crisis within the Indigenous communities land in CHT, Bangladesh

Large-scale Development Projects Numerous development initiatives in the Chittagong Hill Tracts (CHT) have intensified the adverse effects of climate change on the community. Indigenous Elders and Knowledge-keepers have conveyed that these development projects, often supported either forcibly or with direct/indirect governmental and industrial backing, are typically undertaken without adequate information, consultation, or consent from the Indigenous communities. Such large-scale projects encompass hydroelectric dams, tourist areas, profit-oriented monoculture cultivation encroaching upon traditional forests, multinational tobacco plantations, brick industries, external fisheries, commercial roads in hilly areas, and illegal outsider settlements. These human-induced risks pose severe environmental repercussions, disregarding traditional environmental sustainability practices, leading to the depletion of traditional forests, loss of biodiversity, and disruption of waterways. Indigenous communities find themselves marginalized in the decision-making processes related to these projects, intensifying their vulnerability. According to community Elders and Knowledge-keepers, these factors, among others, significantly contribute to the heightened impact of climate change on many Indigenous communities in CHT. They have outlined key climate change impacts on their communities, as depicted in Fig. 2.3.

Declining Drinking and Daily Water Availability A prominent consequence of human-induced climate change is the diminishing water sources crucial for numerous Indigenous communities, especially those reliant on hillside water res-

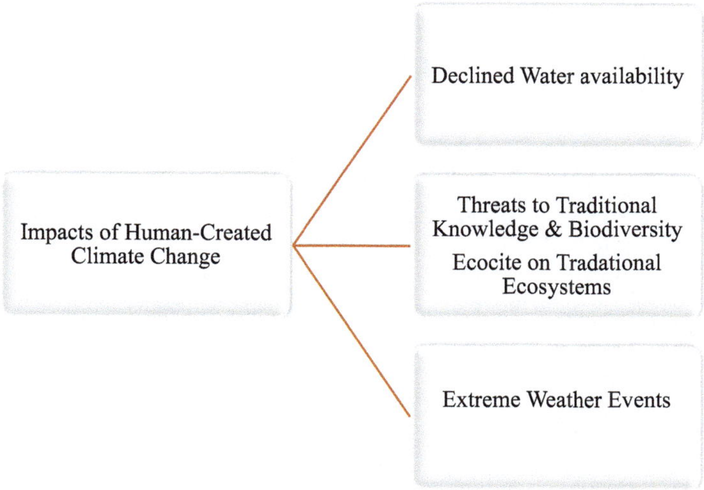

Fig. 2.3 Illustrates the manifold adverse impacts on numerous Indigenous communities resulting from human-induced climate change

ervoirs (see Fig. 2.4). An Elder from the community emphasized, "The human-induced water crisis poses a significant threat to our community, resulting in water scarcity and diminished access to clean water outlets for Indigenous populations." Alterations in traditional forest landscapes have disrupted rainfall patterns, amplifying evaporation rates and contributing to the depletion of rivers, streams, and water reservoirs. Another Elder elaborated on the profound impacts on the community, stating, "The water crisis has severe repercussions on the livelihoods, agriculture, and traditional fishing practices of Indigenous communities." Likewise, a Knowledge-keeper emphasized the sacred role of water in cultural practices, stating, "Water holds spiritual significance in our cultivation rituals and ceremonies, and the current scarcity has plunged us into a severe crisis." Another Elder attributed the water crisis to human activities, noting, "Our local water resources are increasingly imperilled by pollution stemming from industrial practices, including commercial tobacco cultivation, break fields, stone crushing, and construction. This pollution adversely affects both the availability and quality of water, impacting our communities' health, agriculture, and traditional fishing practices."

Rise in Extreme Weather Events Numerous Indigenous Elders and Knowledge-keepers emphasized that climate change has amplified extreme weather events in the Chittagong Hill Tracts (CHT), encompassing floods, cyclones, landslides, and droughts. These events result in the devastation of homes, agricultural lands, and infrastructure, leading to the displacement of communities. Indigenous communities, predominantly residing in remote and marginalized areas, encounter escalated risks and challenges in recovering from the aftermath of these events. As one Elder expressed,

Fig. 2.4 Human-created water crisis impact on the community. This visual representation illustrates the adverse effects of the human-induced water crisis on the community

> The predictability of rainfall patterns in the CHT has diminished. Previously, we could forecast the impact of rain, anticipate patterns, and understand how they influenced the timing and distribution of precipitation. However, now the rainfall has become erratic and unpredictable. This unpredictability adversely affects our traditional agricultural practices, including planting and harvesting seasons, which are vital for our communities subsistence farming. It undermines our traditional practices, and food security, and disrupts the traditional land-based knowledge associated with sustainable cultivation.

Impact on Biodiversity and Traditional Knowledge

Climate change poses a significant threat to the diverse traditional biodiversity of the Chittagong Hill Tracts (CHT) and the accompanying traditional knowledge systems upheld by Indigenous communities. An Indigenous Elder elucidated, "The human-created climate crisis has severely impacted our historical traditional knowledge cultivation system. Our cultivation is integral to the strength of our land and ecosystem. However, current profit-oriented plantation projects, such as monoculture plantations like mango, banana, and lumber, have brought about detrimental changes in the distribution and abundance of traditional native plants and animal species. These new profit-driven plantations pose potential threats to traditional practices like gathering medicinal plants, hunting, and traditional seed saving. The loss of biodiversity undermines the resilience and cultural heritage of Indigenous

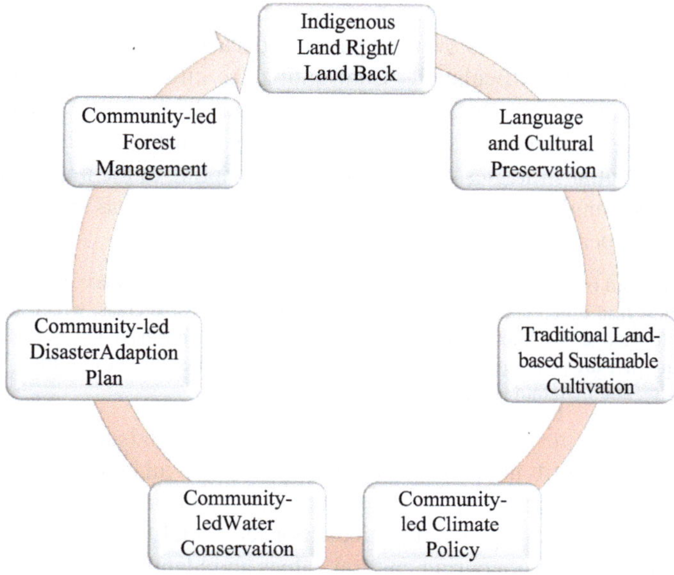

Fig. 2.5 Shows community-engaged and community-led climate change solutions

communities." Another Elder lamented, "The new forms of cultivation and forestation represent the ecocide of our traditional forest." He further said,

> We are deeply connected with our traditional ecosystems [indicating traditional plants, insects, animals, water], which are crucial for our sustainable livelihoods and cultural practices. Many human-led activities on our land pose severe risks to our traditional sustainable ecosystems, including changes in species composition, increased forest fires, and the spread of invasive species. These changes disrupt our communities' reliance on forest resources and traditional knowledge systems.

Effectively addressing the impacts of human-created climate change on Indigenous communities in the Chittagong Hill Tracts (CHT) necessitates community-engaged and community-led approaches. These approaches involve recognizing and respecting Indigenous rights, supporting adaptation strategies, integrating traditional knowledge systems into climate change policies, and promoting sustainable land and resource management practices. Scientific evidence supports the efficacy of community-engaged and community-led approaches in empowering communities. These approaches enhance resilience and ensure meaningful participation in decision-making processes, as illustrated in Fig. 2.5.

Recognizing Indigenous Land Rights

The significance of Indigenous land rights emerges as a critical factor in effectively addressing climate change in the Chittagong Hill Tracts (CHT), as emphasized by Indigenous Elders and Knowledge-keepers. Many Elders have gone so far as to call for the return of lands from governments, illegal settlers, and industries that have forcibly taken their territories over the past five decades. Stressing the vital role of land rights, one Elder articulated, "Confronting the climate change crisis in our communities hinges on our land rights. The acknowledgment and respect for Indigenous peoples' land rights are not only matters of justice and human rights for us but also essential strategies for mitigating and adapting to the impacts of climate change across Bangladesh as a whole." A similar sentiment was shared by another Knowledge-keeper,

> Indigenous land rights are essential in dealing with human-created climate change crises as our peoples' traditional lands often encompass significant forested areas, which are critical for climate change. We know that our community-managed lands have lower deforestation rates than other areas. Respecting our land rights can help prevent deforestation, preserve biodiversity, and maintain these critical crises.

The significance of respecting Indigenous land rights extends to safeguarding cultural heritage and fortifying community resilience, as articulated by numerous community members. Upholding their traditional territories enables Indigenous communities to sustain age-old practices, bolster social unity, and navigate the evolving climate while honouring cultural values and identities.

Traditional Language and Cultural Preservation

Preserving traditional knowledge and language is paramount in addressing the climate crisis within Indigenous communities in the Chittagong Hill Tracts (CHT). An Elder underscored, "Indigenous community-led traditional land-water management practices are intricately linked to Indigenous communities' cultural identity and traditional knowledge systems. These practices, handed down through generations, nurture a robust sense of community cohesion and facilitate the intergenerational transmission of knowledge." Consequently, safeguarding and promoting Indigenous languages and traditional practices emerge as pivotal elements for Indigenous peoples' climate resilience.

Traditional Land-based Cultivation

Indigenous land-based cultivation became a significant way to deal with climate change (Fig. 2.6). One of the Indigenous educators said,

> our community-led traditional cultivation system has provided much scientific evidence regarding how to live sustainably with our environment. We have evidence from hundred years. Our community-led and community-engaged traditional cultivation is critical to adapt to our local topography and climate. Our methods often involve terracing, shifting cultivation, and agroforestry, which helps to retain soil moisture, prevent erosion, and ensure sustainable agricultural production. However, current Western [indicating

> Bangladeshi mainstream climate change] approaches have many things to learn from traditional Indigenous approaches to create meaningful collaboration to address the human-created climate change crises.

Community-Driven Strategies

Numerous community Elders, Knowledge-keepers, and educators advocate for a community-led approach to addressing climate change crises. One Elder said, "Our collective wisdom positions us to tackle climate challenges through community-driven initiatives." Similarly, one of the Elder said,

> Many of our Indigenous communities are often on the front lines of climate change impacts due to our close dependence on natural resources for our livelihoods. We have spiritual relationships in our lands. Our people deeply understand local water resources and our conservation needs. We have developed traditional systems to manage water, including the construction of ponds, reservoirs, and irrigation channels. These systems enable efficient water use, especially during periods of low rainfall, and contribute to sustaining cultivation and livelihoods.

A knowledge-keeper highlighted the effectiveness of a community-led approach in managing their traditional forests, stating, "Our forests play a crucial role in the livelihoods and cultural traditions of our communities in the CHT. Practices such as community-based forest management, rooted in traditional land-water management, serve to safeguard and conserve our forests." Hence, community-driven

Fig. 2.6 Shows indigenous traditional cultivation and sustainable agriculture

strategies, guided by Indigenous land-based knowledge, prove efficient in climate change solutions, encompassing sustainable harvesting, selective logging, and reforestation to uphold ecological balance and biodiversity protection.

Community-Led Disaster Adaptation Strategies

The Chittagong Hill Tracts (CHT) are susceptible to natural disasters, such as landslides and flash floods. Numerous Indigenous Elders and Knowledge-keepers underscored that Indigenous communities have devised traditional land-water management systems, led by the community, to alleviate the impact of such disasters. Initiatives like constructing check dams and contour bunds effectively reduce soil erosion and prevent landslides. Moreover, Indigenous knowledge is critical in selecting suitable settlement sites to minimize risks.

Learning Reflections This research focuses on the indispensability of Indigenous community-led approaches in addressing the human-created climate change crisis in CHT and Bangladesh (Callaghan et al., 2020; Mehran et al., 2017; Steffen et al., 2011). The challenges posed by human-induced climate change demand solutions driven by the communities themselves, as Indigenous groups possess invaluable knowledge, experiences, and practices that can significantly contribute to effective adaptation and mitigation efforts (Mehran et al., 2017; Steffen et al., 2011). Indigenous communities' deep understanding of their local ecosystems, weather patterns, and climate change impacts on their surroundings positions them as vital contributors to sustainable resource management and adaptation strategies.

Indigenous community-led solutions embody the principles of self-determination and empowerment by acknowledging that Indigenous communities should determine their futures (Kipp et al., 2019; Vogel et al., 2022). With a profound connection to their lands, cultures, and ways of life, involving Indigenous communities as active participants and decision-makers fosters a sense of ownership and agency (Cameron et al., 2021; Maldonado et al., 2021). This approach empowers communities to design and implement culturally appropriate, sustainable, and effective solutions tailored to their unique challenges. Collaboration with Indigenous communities as equal partners in climate change initiatives ensures the integration of their rights, knowledge, and perspectives into policy development, planning, and implementation processes (Bullock et al., 2023; Zurba & Papadopoulos, 2023). Indigenous communities in the Chittagong Hill Tracts call for action from the Bangladesh government, policymakers, and international organizations to recognize and safeguard Indigenous land rights, aligning with international frameworks such as the United Nations Declaration on the Rights of Indigenous Peoples (UNDRIP) (Claeys & Edelman, 2020).

Community-led solutions to human-created climate change challenges demonstrate reverence for Indigenous communities' knowledge, experiences, and rights (Datta, 2023, 2024; Trott et al., 2023). By centering Indigenous knowledge and practices, empowering communities, promoting resilience and sustainability, preserving cultural continuity, and fostering collaboration, these solutions effectively address the distinct climate change impacts Indigenous communities face. Upholding community-led approaches can lead Bangladesh towards more equitable and sustainable practices in climate change mitigation, adaptation, and resilience, benefiting Indigenous communities and contributing to global efforts to address the climate crisis. The impacts of human-created climate change on Indigenous communities in the CHT underscore the urgent need for community-led solutions (Farley, 2022; Tu'itahi et al., 2021). This research has illuminated the intricate relationship between human activities and environmental challenges faced by Indigenous peoples. The transformative research conducted within these communities has revealed that traditional knowledge, cultural practices, and land-based approaches are crucial elements for addressing the multifaceted impacts of climate change.

Community-led approaches cannot be overstated, as they empower Indigenous communities to harness their invaluable knowledge and experiences in developing effective adaptation and mitigation strategies (Justrabo, 2023; Quinn et al., 2022). This research has demonstrated that Indigenous communities deeply understand their local ecosystems, weather patterns, and the specific impacts of climate change on their surroundings. Integrating this traditional knowledge into community-led solutions ensures sustainability and respects the principles of self-determination and empowerment. Acknowledging and supporting Indigenous community-led initiatives becomes paramount as Bangladesh and the global community grapple with the escalating climate crisis (Ali et al., 2021; Muthukrishnan & Datta, 2023). The lessons learned from Indigenous communities in the CHT emphasize the importance of recognizing and protecting Indigenous land rights, preserving cultural heritage, and fostering sustainable practices. By embracing community-led solutions, we pave the way for a more equitable and resilient future, where Indigenous communities play a central role in shaping the path towards climate resilience and environmental sustainability.

Chapter 3
Human-Created Climate Crisis and the Significance of Land-Based Adaptation within and from Munda Indigenous Communities in Coastal Areas in Bangladesh

This chapter explores the meaningful implications of Indigenous knowledge systems to deal with human-created climate crisis adaptation strategies, focusing on land-based practice, Indigenous sovereignty, and youth-led initiatives. The chapter examines how the Monda Indigenous communities in the coastal area are struggling with the human-created climate crisis such as land grabbing, displacements, profit-oriented fish farming, clearing natural forests for profit, and only western forest management by the mainstream people and their profit. This chapter also highlights how the traditional land-based knowledge and spirituality approach—a critical framework that brings Indigenous communities self-determination and climate justice—enhances climate resilience. Through a focus on intergenerational land-based learning, the chapter highlights the importance of land-based culture and practice in protecting cultural practices and creating environmental stewardship across generations. It also addresses the central role of Indigenous sovereignty in climate governance, where communities assert control over their land as part of climate adaptation strategies. The chapter explores how climate justice frameworks, informed by decolonial perspectives, can address the disproportionate impacts of climate change on marginalized communities, particularly Monda Indigenous peoples. This chapter aims to provide a comprehensive understanding of how Indigenous knowledge and governance systems offer sustainable solutions to human-created climate crises, promoting both cultural revitalization and climate resilience.

The rising climate crisis, largely human-induced, has emerged as one of the most pressing global challenges (IPCC, 2021). Anthropogenic factors such as industrial emissions, deforestation, and unsustainable land-use practices have accelerated climate change, disproportionately affecting vulnerable communities worldwide (Steffen et al., 2015). In Bangladesh, the impacts are starkly evident, particularly in coastal regions inhabited by Indigenous and minority groups such as the Munda communities (Rahman et al., 2020). Despite contributing minimally to global carbon emissions, these communities bear the brunt of rising sea levels, increasing salinity, and extreme weather events (Islam et al., 2019). However, the voices and

R. Datta, *Decolonizing Climate Adaptation*,
https://doi.org/10.1007/978-981-96-9140-1_3

experiences of Indigenous and minority communities remain underrepresented in climate discourse and policymaking (Whyte, 2018).

One critical research gap lies in understanding how much of the current climate crisis is directly human-created and how this has exacerbated vulnerabilities in both Bangladesh and beyond. Scientific evidence underscores that the majority of climate-related changes—including intensified cyclones, altered precipitation patterns, and environmental degradation—are driven by human activities (Hansen et al., 2013; NASA, 2021). Industrial activities, fossil fuel consumption, and deforestation have not only increased atmospheric carbon dioxide levels but also destabilized ecosystems critical for climate regulation (Steffen et al., 2015). Yet, there is limited research examining the localized impacts of these phenomena on Indigenous communities who have historically lived in harmony with their environments (Ford et al., 2020). This omission obscures the critical insights Indigenous land-based perspectives can offer for climate adaptation and mitigation strategies (Smith & Sharp, 2012).

Indigenous knowledge systems, rooted in centuries of environmental stewardship, offer a wealth of adaptive practices and sustainable solutions (Berkes, 2009). The Munda communities in coastal Bangladesh exemplify how traditional ecological knowledge can contribute to resilience against climate-related adversities. Rituals such as Mansa Puja, which honors nature deities for protection, and resource conservation practices like rainwater harvesting reflect a profound understanding of their environment (Rahman & Mallick, 2022). These practices are not merely cultural artifacts but are adaptive strategies grounded in lived experience and collective memory. For instance, rainwater harvesting has become an essential response to salinity intrusion in freshwater sources, ensuring access to potable water despite environmental challenges (Ahmed, Haque, & Neelormi, 2019). Recognizing and integrating these perspectives is not only a matter of justice but also a pragmatic approach to addressing climate vulnerabilities effectively (Whyte, 2020).

However, the exclusion of Indigenous insights from mainstream adaptation frameworks represents a significant gap in the pursuit of equitable and sustainable climate responses (Kothari et al., 2019). Global climate strategies often prioritize technological and economic solutions while overlooking the localized, community-driven approaches that have sustained Indigenous populations for centuries (Gupta & Sharma, 2020). The Munda's experiences highlight the importance of incorporating land-based knowledge into adaptation and mitigation policies. By valuing these contributions, policymakers can develop responses that are not only inclusive but also more effective in addressing the complex realities of climate change (Ford et al., 2020).

Bangladesh's unique geographic and socioeconomic context further underscores the urgency of this research. As a low-lying deltaic nation, it is particularly vulnerable to climate-induced phenomena such as sea-level rise, cyclones, and flooding (Islam et al., 2019). These challenges disproportionately impact coastal and Indigenous communities, who often lack the resources and infrastructure needed to adapt. Yet, global discussions on climate finance and resilience frequently neglect the needs and contributions of such marginalized groups, perpetuating inequities in climate governance (Adger et al., 2014). The objectives of this chapter are as follows:

- To investigate the disproportionate impacts of the human-created climate crisis on the Munda Indigenous communities in Bangladesh (Ahmed, Mallick, & Rahman, 2019).
- To explore how Indigenous land-based knowledge systems contribute to resilience and adaptation from within these communities (Roy & Chan, 2019).
- To advocate for the inclusion of Indigenous perspectives in climate policies and adaptation strategies at both local and global levels (Datta, 2022).

By addressing these objectives, this paper seeks to bridge critical research gaps while emphasizing the importance of justice and equity in climate adaptation. Acknowledging the human causation of climate crises necessitates a re-evaluation of existing paradigms that have historically marginalized Indigenous voices (Whyte, 2018). Furthermore, leveraging the resilience inherent in Indigenous knowledge systems provides a path toward sustainable and inclusive climate governance (Crate & Nuttall, 2009).

The urgency of this work cannot be overstated. As the climate crisis accelerates, the need for culturally sensitive and community-driven adaptation strategies becomes increasingly apparent (Gupta & Sharma, 2020). While technological innovations and large-scale interventions remain important, they must be complemented by localized approaches that empower marginalized communities to lead their own resilience efforts. The Munda's practices, from conserving freshwater ponds to adapting to shifting agricultural patterns, offer invaluable lessons for global climate strategies (Rahman et al., 2020).

The subsequent sections of this chapter focus on the multifaceted impacts of the climate crisis on Munda communities, from environmental degradation and social marginalization to systemic failures. These sections also emphasize the resilience embedded in Indigenous practices, underscoring the urgency of integrating these insights into broader climate adaptation policies (Watson et al., 2021). Drawing on empirical findings and theoretical frameworks, this research contributes to bridging the gap between global climate narratives and localized, Indigenous-led solutions. Ultimately, this paper underscores that fostering resilience and justice in the face of climate change necessitates a paradigm shift toward inclusive and land-based adaptation strategies (Ford et al., 2020).

Moreover, the findings aim to inspire actionable recommendations for policymakers, researchers, and practitioners (Table 3.1). By highlighting the lived experiences of the Munda communities, this paper advocates for a participatory approach to climate governance that centers on the voices of those most affected. Such an approach aligns with emerging global frameworks that emphasize equity and inclusion as foundational principles for effective climate action (Adger et al., 2014). Through this lens, the research not only addresses immediate vulnerabilities but also contributes to the broader goal of climate justice for Indigenous and marginalized communities worldwide.

Table 3.1 Highlights the primary themes from your research, with their associated subthemes. This format can assist in organizing the content for analysis and comparison, providing a clear overview of how each theme relates to broader research areas in Indigenous knowledge, environmental governance, and climate justice

Main theme	Subtheme(s)
Indigenous knowledge systems	Integration of *two-eyed seeing*
	Incorporating traditional knowledge and western science
Land-based education	Youth intergenerational learning
	Preservation of cultural knowledge through land-based practices
Environmental governance	Indigenous sovereignty over water and energy resources
	Traditional governance systems in climate adaptation
Climate justice and anti-racism	Addressing social injustice and climate change
	Dealing with racism and colonialism in environmental policy
Youth-led climate change education	Empowering indigenous youth through land-based climate education
	Role of youth in fostering climate resilience in communities

Findings

The human-created climate crisis has significant impacts on Monda Indigenous communities, but its consequences are disproportionately severe for Indigenous communities in vulnerable coastal regions. In Bangladesh, the Munda Indigenous communities face unique challenges stemming from environmental degradation, loss of livelihoods, and systemic marginalization. These challenges are compounded by their geographical exposure to rising sea levels, frequent cyclones, and increasing salinity. Yet, the resilience of these communities' manifests through their ancestral knowledge, sustainable practices, and cultural cohesion. This paper explores the themes of environmental degradation, Indigenous land-based adaptation strategies, social marginalization, health and well-being impacts, and systemic failures to emphasize the urgent need for inclusive climate policies.

Environmental Degradation and Its Impacts Environmental degradation poses a significant threat to coastal communities worldwide, and for the Munda Indigenous people of Bangladesh, this reality is particularly acute. The interplay of rising sea levels, saline intrusion, and frequent natural disasters has created a cascade of challenges that undermine the region's ecological balance and the community's way of life. This theme explores the severe impacts of environmental degradation, focusing on increased salinity, the destruction of livelihoods and land, and the issue of forced migration and shelter insecurity. Through the voices and lived experiences of the Munda, the urgency for targeted climate interventions becomes starkly evident.

Increased Salinity and Environmental Degradation The coastal regions of Bangladesh, including Munda communities, are severely affected by increasing salinity. This environmental degradation compromises their water, food, and health security. Testimonies highlight that "everywhere is salinity. We store the water from the rain and drink that way," underscoring the dire reliance on rainwater amidst the unavailability of freshwater sources. The pervasive salinity not only affects drinking water but also renders agricultural land unproductive, leading to a significant reduction in crop yield. Skin diseases and other health issues arising from contact with saline water further exacerbate the community's vulnerability. The inability to grow vegetables or sustain traditional farming practices reflects a stark disruption of livelihoods and food security. Salinity intrusion is not merely an environmental challenge but a direct threat to the survival of Munda communities.

Destruction of Livelihoods and Land Environmental disasters, exacerbated by the climate crisis, disrupt livelihoods and devastate infrastructure. Accounts such as "our house gets broken; chickens and cattle are washed away" and "the road is broken, so we cannot reach the hospital, especially for pregnant women" reflect the cascading effects on safety, economy, and healthcare accessibility. The destruction of roads, homes, and agricultural lands highlights the fragility of infrastructure in these coastal regions. Livelihoods centered around farming, fishing, and livestock are frequently destroyed, leaving families without income or food. The economic losses incurred during each disaster create a cycle of poverty that is difficult to escape. The displacement caused by the destruction of homes further strains the community's resilience, as temporary shelters often lack basic amenities and security. For pregnant women and the elderly, the situation is particularly dire due to the absence of accessible healthcare facilities.

Forced Migration and Shelter Issues Frequent natural disasters force Munda families to relocate to temporary shelters, often leading to displacement and insecurity. Despite the challenges, the resilience of these communities is evident in statements like, "we stay here, no matter if we live or die." This determination to remain on ancestral land underscores their deep connection to place and heritage. However, the lack of adequate cyclone centers and the overcrowding of existing facilities often leave families exposed to the full impact of storms. The absence of secure land tenure further complicates their situation, as many families live on land they do not own, making relocation and rebuilding efforts even more challenging. Forced migration disrupts community cohesion and exposes families to new risks, including exploitation and loss of cultural identity.

The environmental challenges faced by the Munda communities reflect a broader struggle against the compounded effects of climate change and systemic neglect. Increased salinity, destruction of livelihoods, and forced migration not only threaten their physical survival but also erode their cultural identity and social cohesion. Addressing these issues requires a multidimensional approach that integrates environmental restoration, infrastructure development, and land security. By prioritizing the voices and needs of the Munda, policymakers can design interventions that are both culturally sensitive and ecologically sustainable, ensuring that these resilient communities can continue to thrive despite the mounting challenges.

Indigenous Knowledge and Land-Based Adaptation Indigenous knowledge systems have long been pivotal in fostering resilience and adaptation to environmental challenges. For the Munda communities in coastal Bangladesh, their ancestral practices, sustainable resource management, and collective memory serve as vital tools in combating the impacts of the human-created climate crisis. This theme delves into the cultural and practical dimensions of Indigenous knowledge, highlighting how rituals, conservation strategies, and historical lessons inform their adaptive capacities. By foregrounding these practices, we can better understand the symbiotic relationship between cultural heritage and environmental sustainability.

Ancestral Practices and Rituals Munda communities rely on ancestral practices and rituals as adaptive strategies. Celebrating Mansa Puja and invoking Maa Mansa during travels for safety and protection signify their deep spiritual connection to nature. These rituals play a critical role in maintaining psychological resilience during crises. They also serve as a reminder of the community's historical experiences with adversity and the ways in which cultural practices have been used to navigate challenges. Such rituals foster a sense of unity and purpose, providing emotional support and a collective framework for action. By honoring their spiritual traditions, the Munda reinforce their identity and resilience in the face of environmental adversity.

Resource Conservation and Sustainable Practices Rainwater harvesting and maintaining freshwater ponds demonstrate the Munda's sustainable water management practices. Statements like "we do not pollute the pond that contains fresh water" illustrate their commitment to environmental stewardship. These practices are deeply rooted in Indigenous knowledge systems that prioritize sustainability and balance with nature. By using rainwater for drinking and maintaining strict rules against polluting freshwater sources, the Munda ensure the longevity of their vital resources. Such practices not only provide immediate benefits but also offer valuable lessons for broader climate adaptation efforts. Their approach to resource management reflects a harmonious relationship with the environment, emphasizing the importance of conserving natural assets for future generations.

Collective Memory of Past Adaptations Memories of surviving past disasters, such as "during Aila, we used to stay at the school and also on the road," provide valuable lessons for community resilience. These collective memories serve as a repository of strategies and actions that have proven effective in mitigating the impacts of disasters. They also strengthen social bonds by fostering a shared sense of purpose and identity. By recalling and building upon past experiences, the Munda are able to develop adaptive strategies that are both practical and culturally relevant. This reliance on collective memory highlights the importance of preserving Indigenous knowledge as a critical component of climate resilience.

In this theme, the Munda communities' explained their reliance on Indigenous knowledge underscores the value of integrating cultural wisdom into contemporary climate adaptation strategies. Their ancestral practices, sustainable resource management, and collective memory not only enhance their resilience but also offer lessons for addressing broader environmental challenges. Recognizing and respecting these knowledge systems is crucial for fostering inclusive and effective climate solutions. By supporting the preservation and application of Indigenous practices, policymakers and stakeholders can empower the Munda and similar communities to thrive amidst the growing threats posed by climate change.

Health and Well-Being Impacts The human-created climate crisis has far-reaching effects on the health and well-being of vulnerable communities. For the Munda Indigenous people of coastal Bangladesh, the consequences are particularly dire. From waterborne diseases to limited healthcare access and the psychological toll of environmental instability, these impacts exacerbate existing vulnerabilities and threaten overall resilience. This theme examines the intersection of health challenges and climate change, emphasizing the urgent need for comprehensive strategies to address the physical, mental, and systemic health inequities faced by the Munda.

Waterborne Diseases and Health Issues Contaminated water sources significantly affect the health of Munda communities, leading to widespread cases of diarrhea, skin diseases, and other waterborne illnesses. The reliance on saline water, particularly during menstruation, intensifies health challenges for women, creating both physical and psychological burdens. "During menstruation, we face trouble using saline water" encapsulates the compounded difficulties women endure in maintaining hygiene and dignity. These health issues are aggravated by the lack of clean drinking water and inadequate sanitation facilities, which are critical for preventing disease outbreaks. Ensuring access to safe and clean water is essential for improving public health outcomes in these vulnerable areas.

Lack of Healthcare Access Healthcare facilities in Munda regions are often inaccessible or insufficiently equipped to meet the needs of the community. Testimonies reveal a heavy reliance on local quacks or distant government health centers, which limits timely medical intervention. "If someone gets sick at night, the quack won't open the door for us" illustrates the precarious nature of healthcare access and the lack of trust in available services. The financial burden of seeking medical care further complicates the issue, as families must often prioritize immediate survival needs over health expenses. Expanding healthcare infrastructure and providing culturally sensitive medical services are critical steps toward addressing these disparities.

Psychological Impacts of Disaster The loss of land and frequent environmental disasters take a profound toll on the mental well-being of Munda communities. The uncertainty of living in disaster-prone areas creates a constant state of anxiety and worry. Statements like "people become worried and get sick" and "I feel bad because we don't have our own land" reflect the psychological strain caused by systemic insecurity and environmental instability. The erosion of traditional lifestyles and the loss of cultural identity further exacerbate these mental health challenges. Incorporating mental health support into disaster response frameworks and fostering community-based resilience practices can help mitigate the psychological impacts of climate change.

The health and well-being of Munda communities are deeply intertwined with the effects of the climate crisis. Waterborne diseases, inadequate healthcare access, and the psychological toll of environmental instability highlight the urgent need for targeted interventions. By improving water and sanitation infrastructure, expanding healthcare services, and addressing mental health challenges, policymakers can enhance the resilience and quality of life for the Munda. Integrating health considerations into climate adaptation strategies is essential to ensure a holistic and equitable response to the ongoing crisis.

Systemic Failures and Inequalities Systemic failures and inequalities exacerbate the vulnerabilities of Munda communities, undermining their resilience and adaptation efforts in the face of the climate crisis. Corruption, infrastructure deficiencies, and insecure land tenure reflect the broader neglect and marginalization experienced by Indigenous groups. This theme explores how these systemic issues compound the challenges posed by environmental degradation, emphasizing the need for structural reforms and inclusive governance to address these inequities.

Corruption and Inefficient Relief Distribution Corruption and mismanagement of relief resources exacerbate the hardships faced by Munda communities. Local authorities and intermediaries often exploit disaster relief systems, diverting resources meant for vulnerable populations. Statements such as "the member and the chairman are taking everything" highlight the systemic exclusion of marginalized groups from government aid. This corruption not only denies communities the support they desperately need but also erodes trust in public institutions. Establishing transparent mechanisms for relief distribution and involving community representatives in decision-making processes can address these inequities and improve disaster response efficacy.

Infrastructure Deficiencies Inadequate infrastructure isolates Munda communities, limiting their access to essential services and opportunities. Poor road conditions, as described in "to the nearby villages where all the mainstream people are living, they have a good road, but we don't have," reflect the disparity in development priorities. These deficiencies impact education, healthcare, and economic activities, further entrenching cycles of poverty. For instance, children often face disruptions in their education during floods due to inaccessible schools. Improving infrastructure in marginalized areas, including roads, schools, and healthcare facilities, is essential for fostering equitable development and enhancing community resilience.

Land Ownership and Security The lack of secure land ownership leaves Munda families in a state of perpetual vulnerability. Living on land they do not own exposes them to the constant threat of displacement and limits their ability to invest in long-term improvements. "If we had our own land, it would be better" reflects the community's aspiration for stability and empowerment through land rights. Secure land tenure not only provides a sense of belonging and safety but also enables communities to plan for the future, invest in sustainable practices, and withstand the impacts of climate change. Advocating for land rights and integrating land security into climate adaptation policies are critical for empowering Indigenous communities.

Addressing systemic failures and inequalities is essential to building resilience among Munda communities. By tackling corruption, investing in infrastructure, and ensuring land ownership rights, policymakers can create a more equitable and sustainable environment for Indigenous populations. Structural reforms must prioritize the needs and voices of marginalized groups to dismantle the barriers that perpetuate vulnerability and exclusion. Through inclusive and transparent governance, it is possible to foster resilience and empower the Munda to navigate the challenges of a changing climate.

Discussion

This research investigates key aspects of Indigenous knowledge systems and their role in addressing climate change and promoting sustainability. All the above themes and sub-themes from our finding sections showcase the experiences of Munda Indigenous communities in coastal Bangladesh and expose the multifaceted impacts of the human-created climate crisis. From environmental degradation and health vulnerabilities to systemic inequalities and cultural resilience, their narratives underscore the urgency of integrating Indigenous knowledge into climate adaptation strategies. Sustainable development must prioritize the rights and well-being of marginalized communities like the Munda to foster equitable and effective responses to the climate crisis. Recognizing their ancestral practices, addressing systemic barriers, and ensuring land security are essential steps toward sustainable and inclusive adaptation. By valuing and supporting land-based solutions, we can co-create a future where resilience and justice prevail. The human-created climate crisis has profound impacts worldwide, but its consequences are disproportionately severe for Indigenous communities in vulnerable coastal regions. In Bangladesh, the Munda Indigenous communities face unique challenges stemming from environmental degradation, loss of livelihoods, and systemic marginalization. These challenges are compounded by their geographical exposure to rising sea levels, frequent cyclones, and increasing salinity. Yet, the resilience of these communities manifests through their ancestral knowledge, sustainable practices, and cultural cohesion. This paper explores the themes of environmental degradation, Indigenous land-based adaptation strategies, social marginalization, health and well-being impacts, and systemic failures to emphasize the urgent need for inclusive climate policies.

With a focus on the intersections of Indigenous sovereignty, land-based education, and climate action, this research brings forward both contemporary and traditional approaches to climate resilience in Indigenous communities. The findings of this study build upon and expand current understandings of climate change education, water, and energy sovereignty, and the role of intergenerational learning in Indigenous contexts. This discussion compares the findings from this study with existing literature on related topics, including the integration of Indigenous and Western knowledge systems, youth-led initiatives, and land-based education. The work also references broader issues surrounding climate justice, anti-racist and decolonial frameworks, and Indigenous sovereignty in the context of environmental governance.

The Importance of Indigenous Land-Based Knowledge One of the primary findings of this study is the importance of Indigenous land-based knowledge and practice to deal with human-created climate crisis in Monda Indigenous communities, a framework that centers Indigenous knowledge. The findings echo the work of Henry and Pene (2001) in their exploration of Kaupapa Māori, where they argue for the respect recognition and validation of Indigenous epistemologies. Our study supports this argument by demonstrating how communities that incorporate both knowledge systems are better equipped to address complex issues like climate change and environmental degradation (Bartlett et al., 2012; Hill, 2019). The application of Indigenous land-based knowledge and practice allows for a deeper understanding of climate risks and environmental stewardship, as it acknowledges the interconnectedness of land, water, and energy systems in Indigenous contexts (Shaw, 2017). Similarly, other studies on Indigenous knowledge and climate change adaptation emphasize the significance of integrating traditional knowledge in disaster risk reduction strategies (Ford et al., 2020). Our findings contribute to this body of knowledge by demonstrating how *Two-Eyed Seeing* can foster more inclusive, holistic solutions to environmental challenges.

Land-Based Education and Intergenerational Learning In this chapter, this study highlights the importance of land-based education and intergenerational learning in addressing climate change. This aligns with the findings of numerous scholars who emphasize the role of traditional ecological knowledge (TEK) in climate adaptation strategies (Berkes, 2018; Whyte, 2013). Land-based learning, particularly when led by Elders and community members, serves as a vital means of preserving cultural knowledge while fostering resilience to environmental changes (Simpson, 2014). The intergenerational aspect of land-based education, which is central to our study, supports the idea that knowledge transmission across generations can build community resilience. Recent research indicates that youth involvement in climate change education is crucial for long-term sustainability (Corntassel, 2012). Our findings build upon this by showing that Indigenous youth, when empowered through intergenerational learning, play a significant role in climate change adaptation and governance. This is consistent with work by scholars like Kimmerer (2013), who has emphasized the crucial role of youth in revitalizing cultural practices and ensuring the transmission of Indigenous knowledge for future generations.

Indigenous Sovereignty as Sustainable Climate Governance The study's findings also demonstrate the centrality of Indigenous sovereignty in climate governance. This is particularly evident in the research on water and energy sovereignty, where Indigenous communities are reclaiming control over natural resources to assert their autonomy and self-determination (LaDuke, 2016). Studies by scholars such as Nelson and Wilson (2018) have shown that Indigenous-led environmental governance models, grounded in traditional knowledge systems, offer sustainable solutions to climate change and environmental degradation. These findings echo these sentiments, underscoring the importance of Indigenous-led governance in mitigating climate risks and fostering environmental stewardship. This reflects the work of Coulthard (2014), who argues that the recognition of Indigenous sovereignty is key to addressing both environmental and social injustices. Furthermore, our research shows that Indigenous communities, by reviving traditional knowledge and governance systems, can more effectively manage natural resources and protect their territories against the impacts of climate change.

Climate Justice and Anti-Racist Frameworks The intersectionality of climate change with social justice issues, such as racism, colonialism, and marginalization, is another key finding in our study. This reflects the growing recognition of climate justice as a movement that addresses the disproportionate impacts of climate change on marginalized communities, particularly Indigenous peoples (Shiva, 2016). Our study's emphasis on anti-racist and decolonial perspectives aligns with the work of scholars like Tuck and Yang (2012), who argue that climate justice must include the decolonization of environmental spaces and the dismantling of structures of power that perpetuate environmental harm. In the context of Black, Indigenous, and People of Color (BIPOC) communities, this study finds that fostering resilience and adaptive capacities requires an active engagement with anti-racist frameworks. By centering Indigenous voices and prioritizing Indigenous-led climate solutions, the study contributes to an understanding of how climate justice can be achieved through collective action and solidarity (Dunlap & Brulle, 2015).

Community-Led Climate Change Education Our study also explores youth-led climate change education initiatives in Indigenous communities, particularly in coastal areas of Bangladesh. This component of the research aligns with global trends in climate education, where youth leadership is increasingly recognized as a transformative force in shaping climate action (UNICEF, 2020). The study's findings support the assertion that empowering Indigenous youth through land-based education programs enables them to become agents of change in their communities, fostering both environmental and cultural resilience.

Research by authors like Jasanoff (2018) has highlighted the critical role of young people in shaping the future of climate governance, especially in Indigenous communities. Our study confirms that by engaging youth in land-based, culturally relevant climate education, communities can strengthen their capacity for sustainable environmental management.

The findings from this chapter contribute significantly to the literature on Indigenous knowledge, climate change, and environmental governance. By emphasizing Indigenous land-based education, Indigenous sovereignty, and decolonial community-led approaches to addressing climate change. These findings align with existing literature that highlights the value of traditional land-based knowledge in climate adaptation and the centrality of youth and intergenerational learning in fostering resilience. This chapter adds depth to ongoing discussions about climate justice by emphasizing the role of Indigenous-led governance and the need for decolonization in climate policy. As Indigenous communities continue to lead efforts in climate resilience and sustainability, the lessons learned from this research can guide future initiatives aimed at fostering climate justice for all.

Chapter 4
Monda Indigenous Youth Perspectives on Climate Crisis and Land-Based Adaptions

This chapter aims to illuminate the multifaceted challenges and opportunities faced by Monda Indigenous youth in the context of the climate crisis. By centering their narratives, it seeks to provide a platform for their voices and advocate for culturally grounded, community-driven solutions. The findings presented herein offer valuable insights into how resilience and innovation can emerge from adversity, providing lessons not only for the Monda community but for global efforts to address the climate crisis. By weaving together their lived experiences, aspirations, and strategies, this chapter aspires to contribute to a broader understanding of how Indigenous youth can lead the way in creating a more sustainable and equitable future.

The growing climate crisis has placed unprecedented pressure on Indigenous communities worldwide, particularly on their youth. Among these, the Monda Indigenous youth have emerged as vital narrators of resilience and adaptability, offering unique insights into the complex interplay between environmental degradation, socio-economic inequities, and cultural preservation. Their narratives are a rich tapestry of challenges and possibilities, showcasing both the vulnerabilities they face and the innovative strategies they employ to navigate these difficulties. This chapter aims to center their voices to provide a nuanced understanding of the issues at hand while exploring pathways to empower these youth as agents of transformative change. The intersection of health and environmental challenges is particularly stark for the Monda community. The salinity crisis has far-reaching effects on women's health, leading to malnutrition, menstrual health issues, and unsafe childbirth conditions (Coulthard, 2014; Whyte, 2017). Additionally, the reliance on contaminated water sources for daily activities exacerbates the spread of waterborne diseases, such as skin infections and gastrointestinal illnesses (Smith, 2012; Wilson, 2008). The absence of adequate sanitation facilities compounds these issues, with many families resorting to unsafe hygiene practices that compromise their dignity and health (Dudgeon & Berkes, 2003).

Education is a fundamental right but remains a distant reality for many Monda youth. Geographical inaccessibility, economic constraints, and cultural barriers

R. Datta, *Decolonizing Climate Adaptation*,
https://doi.org/10.1007/978-981-96-9140-1_4

significantly limit their educational opportunities (Battiste, 2002; Coates, 2004). For girls, societal norms that prioritize early marriage over education further curtail their potential. These barriers perpetuate cycles of inequality, preventing the community from leveraging education as a tool for empowerment and development (Hauser, 2018; Kimmerer, 2013). The Monda youth's resilience is tested by social marginalization and cultural discrimination. Mockery of their Indigenous practices and fear of attacks during festivals erode their confidence and sense of belonging. As Ruma Munda notes, "Mainstream communities mock our religion, saying disrespectful things," underscoring the systemic bias that alienates them from broader society (Leichenko & O'Brien, 2008; Pretty & Smith, 2004). Despite these challenges, their commitment to preserving their cultural identity remains unwavering (Coulthard, 2014; Smith, 2012).

The Monda community's deep connection to their environment provides a foundation for innovative solutions. Integrating traditional ecological knowledge with modern scientific practices can create culturally relevant and sustainable adaptation strategies (Berkes, 2009; Nakashima et al., 2012). For instance, restoring mangroves, developing salinity-resistant crops, and implementing community-led water conservation efforts can mitigate the impact of climate change while preserving their ecological heritage (Hauser, 2018; Walker et al., 2004).

Empowering communities to lead their development initiatives is crucial for creating lasting change. Community-based participatory action research (CBPAR) offers a framework for identifying local priorities and co-developing solutions. This approach not only ensures cultural relevance but also fosters a sense of ownership and accountability among community members (Dudgeon & Berkes, 2003; Wilson, 2008).

Education is a powerful tool for transformation. Establishing local schools, providing scholarships, and integrating Indigenous knowledge into curricula can bridge the gap between traditional and modern systems (Battiste, 2002; Kimmerer, 2013). Moreover, nurturing youth leadership through mentorship programs and capacity-building initiatives can empower the next generation to advocate for their rights and lead community development efforts (Coates, 2004; Whyte, 2017).

Addressing gender disparities requires targeted interventions that challenge discriminatory norms and empower women and girls. Initiatives such as vocational training, microfinance programs, and safe spaces for dialogue can help women gain economic independence and participate more actively in decision-making processes (Escobar, 1995; Hauser, 2018). Ensuring access to quality healthcare and education for girls is equally critical to breaking cycles of poverty and inequality (Ford & Pearce, 2010; Leichenko & O'Brien, 2008). Diversifying income sources and promoting sustainable economic models are essential for reducing dependency on exploitative industries. Cooperative businesses, agricultural innovation, and skills training can provide alternative livelihoods that align with the community's cultural and ecological context (Nakashima et al., 2012; Pretty & Smith, 2004). Such efforts can enhance financial resilience while preserving the community's traditional practices (Adger, 2000; Walker et al., 2004).

This chapter explores the perspectives of Monda Indigenous youth on the climate crisis, focusing on their lived experiences, challenges, and resilience strategies. It begins by contextualizing the impact of the climate crisis on Indigenous communities, highlighting the unique insights provided by Monda youth. The chapter first examines the Impact of Climate Change, exploring environmental changes, livelihood challenges, and water scarcity, with an emphasis on how disasters and salinity threaten community sustainability. The Health and Sanitation section discusses how environmental degradation intensifies women's health issues, water contamination, and inadequate sanitation, affecting the dignity and well-being of the community. The Education and Gender segment addresses systemic barriers to education, gender disparities, and early marriage, revealing socio-economic constraints on youth empowerment. The Livelihood and Poverty section highlights limited job opportunities, economic struggles, and seasonal dependency, illustrating the precariousness of the community's economic systems. Finally, the Cultural Identity and Resilience section explores how Indigenous practices sustain cultural identity, the challenges in preserving traditions, and the aspirations of youth to uplift their community (Fig. 4.1). The chapter concludes by emphasizing the critical importance of integrating Monda youth perspectives into policies and practices to foster sustainable climate adaptation and youth self-determination.

Fig. 4.1 Shows Monda Indigenous Youths who want to bring their traditional land-based culture and practice as climate adaption. This photo was taken during our field research in August 2023

Findings

The Monda Indigenous youth have been vocal about their lived experiences and observations of the climate crisis and its repercussions on their communities. Rooted in a close relationship with the land and natural resources, their perspectives provide a nuanced understanding of the impact of environmental changes, livelihood challenges, and water scarcity. These narratives are not only a testament to their resilience but also highlight the urgent need for culturally relevant, land-based adaptation strategies. This section explores three critical themes derived from the experiences of Monda youth, offering an in-depth analysis of the environmental, social, and economic challenges they face while emphasizing the importance of their voices in climate discourse (Table 4.1).

Impact of Climate Change Climate change has profoundly disrupted the lives and livelihoods of the Monda community, altering their environment, economic activities, and access to essential resources. This theme delves into the specific environmental changes, livelihood challenges, and water scarcity issues articulated by Monda youth.

Environmental Changes Environmental degradation, exacerbated by climate change, has been a pressing concern for the Monda Indigenous youth. Kanika Munda highlights the frequency and severity of natural disasters, stating, "Here we have high tide, cyclones, floods; river erosion, soil erosion, etc. People suffer from these disasters." These phenomena disrupt the ecological balance and directly impact the community's stability. Cyclones and floods destroy infrastructure, displace families, and lead to long-term socio-economic vulnerabilities. The relentless nature of these disasters underscores the need for robust, community-led disaster management strategies. The issue of salinity is another pressing environmental challenge. As Kumila Munda notes, "Saline water makes it difficult to live." Rising sea levels and frequent high tides introduce saline water into freshwater systems, impacting agriculture, drinking water, and biodiversity. Sonika Munda's observation, "When the high tide comes, we don't get fresh water; saline water is everywhere," illustrates the daily struggles faced by the community.

Table 4.1 Shows themes and subthemes from the monda indigenous youth perspectives on climate crisis and land-based adaptions

Theme	Subthemes
Impact of climate change	Environmental changes, livelihood challenges, water scarcity
Health and sanitation	Women's health issues, impact of contaminated water, lack of sanitation facilities
Education and gender	Access to education, gender disparity, early marriage
Livelihood and poverty	Limited job opportunities, economic struggles, seasonal dependency
Cultural identity and resilience	Indigenous practices, community challenges, aspirations for change

To mitigate these challenges, land-based adaptation strategies that integrate traditional ecological knowledge with contemporary climate science are crucial. These could include building salinity-resistant agricultural systems, restoring mangroves to buffer against cyclones, and creating community-driven freshwater reservoirs.

Livelihood Challenges Livelihoods in the Monda community are intrinsically linked to the land and natural resources. However, climate change has significantly disrupted these connections, creating economic precarity. As Kanika Munda observes, "We cannot grow vegetables. We cannot even cultivate rice. Everywhere you can see shrimp quarry." The transition from traditional agriculture to shrimp farming, driven by economic pressures and environmental constraints, exemplifies the complex interplay between climate change and local livelihoods. While shrimp farming may offer short-term economic gains, it often exacerbates soil salinity and undermines long-term food security. The inequitable distribution of resources further compounds these challenges. Ruma Munda's statement, "Rich people are taking land on lease and cultivating shrimp. We cannot do that," highlights the socio-economic disparities that restrict the community's access to productive land. This marginalization perpetuates cycles of poverty and reduces opportunities for economic advancement.

Moreover, the reliance on the Sundarbans for sustenance, as noted by Sonika Munda ("People earn less because they depend on the Sundarbans. They don't have options to earn money and send their children to school"), underscores the vulnerability of the community to environmental degradation. Diversifying livelihood options and ensuring equitable resource distribution are essential steps toward building resilience. Community-based participatory action research (CBPAR) could facilitate the development of sustainable economic models that respect cultural and ecological boundaries.

Water Scarcity Water scarcity is a recurring issue that deeply affects the Monda community's quality of life. Kanika Munda laments, "We suffer a lot from the lack of fresh water. We store rainwater and use it." This reliance on rainwater reflects the inadequacy of freshwater resources, exacerbated by rising salinity levels. Rainwater harvesting, while essential, is not a sufficient long-term solution in the face of increasing climate unpredictability.

The salinity's pervasive impact, as noted by Ruma Munda ("Salinity has an impact on drinking water, cultivation, and growing vegetables"), extends beyond drinking water to disrupt agricultural activities and food security. The contamination of traditional freshwater sources, such as ponds, is another significant concern. Sonika Munda's observation, "Our fresh water pond is no more fresh; it is getting salty," captures the gravity of the issue.

Addressing water scarcity requires a multi-faceted approach. Restoring freshwater ecosystems, implementing desalination technologies, and protecting existing water sources through community-driven conservation efforts are critical. Incorporating Indigenous knowledge, such as identifying and preserving naturally occurring freshwater sources, can complement these initiatives.

The impact of climate change on the Monda Indigenous youth highlights the interconnectedness of environmental, social, and economic challenges. Their narratives underscore the urgent need for adaptive strategies that respect cultural practices while addressing systemic inequities. Environmental restoration, sustainable livelihood programs, and innovative water management systems are essential components of a holistic response. By centering the voices of Monda youth, policymakers and researchers can co-create solutions that are both effective and culturally relevant. The lived experiences of Monda Indigenous youth provide a powerful lens through which to understand the multifaceted impacts of climate change. Their insights emphasize the importance of integrating traditional ecological knowledge with modern adaptive strategies. As the climate crisis intensifies, fostering intergenerational and intercultural collaborations will be essential in developing resilient and inclusive solutions. Ultimately, the resilience and wisdom of the Monda youth are not just assets for their community but valuable contributions to the global fight against climate change.

Health and Sanitation The intersection of health and sanitation with environmental and socio-economic challenges significantly shapes the lives of Monda Indigenous youth. Their insights reflect the compounding effects of climate-induced crises and systemic neglect on physical and mental well-being. Women's health, the pervasive impact of contaminated water, and the absence of adequate sanitation facilities are central issues in this context. Understanding and addressing these challenges require culturally sensitive and community-driven approaches. This section explores these subthemes, providing a detailed account of the struggles and resilience demonstrated by the Monda youth in the face of adversity. Health and sanitation are critical dimensions of human dignity and survival. For the Monda Indigenous community, these areas are severely compromised due to environmental changes and inadequate infrastructure. The youth's narratives offer an intimate look at the physical and emotional toll of these conditions, as well as the innovative coping strategies they employ.

Women's Health Issues The health issues faced by women in the Monda community are deeply intertwined with the salinity crisis and lack of access to essential healthcare services. Kanika Munda highlights the profound physical effects, stating, "Women suffer a lot due to salinity. For example, malnutrition, absence of periods, and period cramps." This observation underscores the systemic neglect of women's health, which is exacerbated by environmental stressors such as salinity intrusion in water and soil. Malnutrition, a consequence of limited access to diverse and nutritious food, further compounds these issues.

The impact on menstrual health is particularly troubling. Kumila Munda notes, "I get skin disease and feel trouble during menstruation because of the water," illustrating how contaminated water directly affects women's daily lives and health. The lack of safe, clean water not only leads to infections but also creates psychological distress and stigma around menstruation. Sonika Munda's poignant statement, "We are living in a crisis of life and death; there is no option for safe delivery," highlights the critical lack of maternal healthcare facilities. This absence forces women to endure unsafe childbirth conditions, resulting in preventable maternal and infant mortality. Addressing these challenges requires a multi-faceted approach, including improving access to healthcare, ensuring clean water supply, and promoting community-based health education initiatives.

Impact of Contaminated Water The pervasive issue of water contamination poses significant health risks for the Monda community. As Ruma Munda explains, "Skin diseases spread out because everyone collects water from the same pond for daily activities." This shared dependency on contaminated water sources exacerbates the spread of waterborne illnesses, including skin infections, gastrointestinal diseases, and other health complications.

Efforts to mitigate these risks are often resource-intensive and insufficient. Ruma Munda's attempt to filter water with fitkari ("I try to filter the water with fitkari before using it") demonstrates the community's resilience and ingenuity in addressing these challenges. However, such measures are stopgap solutions and fail to address the root causes of water contamination.

The sudden onset of floods further disrupts access to clean water and hygiene practices. Sonika Munda's remark, "We cannot bathe properly when the sudden flood comes," underscores how climate-induced disasters amplify existing vulnerabilities. Establishing sustainable water management systems and emergency response plans is essential to mitigate the health impacts of contaminated water.

Lack of Sanitation Facilities The absence of adequate sanitation infrastructure is a pervasive issue that compromises the dignity and health of the Monda community. Kanika Munda's observation, "We don't have sanitary toilets here, and this is a big problem for us, especially during disasters," highlights the heightened vulnerabilities during extreme weather events. The lack of sanitary toilets forces individuals to resort to unsafe and unhygienic practices, increasing the risk of diseases.

The challenges are particularly acute during floods, which create overcrowded and unsanitary living conditions. Ruma Munda's statement, "When floods come, we have no separate space for men, women, and children," points to the lack of privacy and safety in emergency shelters. Such conditions disproportionately affect women and children, exposing them to risks of harassment and health complications.

Moreover, the reliance on natural water bodies for hygiene, as noted by Sonika Munda ("We don't have a bathroom; we take our bath in the nearby ponds"), underscores the dire state of sanitation infrastructure. Developing community-led sanitation programs that include building eco-friendly toilets, creating separate spaces for genders in shelters, and educating the community on hygiene practices are critical steps toward addressing these issues.

The health and sanitation challenges faced by the Monda Indigenous youth are deeply rooted in systemic inequities and environmental degradation. Their narratives highlight the urgent need for targeted interventions that address women's health, mitigate the impact of contaminated water, and improve sanitation facilities.

Collaborative efforts that integrate traditional knowledge, community participation, and sustainable development practices can pave the way for long-term solutions. By amplifying the voices of Monda youth, stakeholders can ensure that these initiatives are inclusive and culturally relevant. The experiences of the Monda Indigenous youth underscore the critical need for a holistic approach to health and sanitation challenges. Their resilience and innovative coping mechanisms offer valuable insights into community-driven solutions. Moving forward, integrating their perspectives into broader health and sanitation policies can create more effective and sustainable outcomes. By prioritizing the well-being of vulnerable populations, we can build a future where health and dignity are accessible to all, even in the face of climate-induced adversities.

Education and Gender Education is a fundamental human right and a cornerstone for social and economic development. For the Monda Indigenous community, however, access to quality education is fraught with challenges, particularly for girls and women. Gender norms, economic constraints, and systemic barriers hinder equitable access to education and perpetuate cycles of inequality. The perspectives of Monda Indigenous youth on education and gender provide valuable insights into these issues, emphasizing the interplay between access to education, gender disparity, and the prevalence of early marriage. This section explores these subthemes, offering a nuanced understanding of the challenges and potential solutions. Education and gender are deeply interconnected dimensions of the Monda Indigenous community's lived experiences. The youth's narratives shed light on the systemic inequities they face, while also highlighting the resilience and agency of individuals striving to create a better future for themselves and their community.

Access to Education Access to education remains a significant challenge for the Monda community due to geographical, economic, and infrastructural barriers. Kanika Munda, who studies at Mohosin Degree College, explains, "I study at Mohosin Degree College, which is very far from here. I go by bus and van." The long distances to educational institutions impose both financial and physical burdens, particularly on families with limited resources. This challenge disproportionately affects girls, as safety concerns and societal expectations often limit their mobility. Primary education is similarly affected by geographic inaccessibility. Kumila Munda notes, "The primary school is so far that children cannot go alone; someone needs to accompany them." This requirement often prevents children from attending school regularly, as caregivers must prioritize other responsibilities over accompanying children to distant schools. Economic constraints further exacerbate these barriers. Sonika Munda's statement, "We cannot send our children outside for higher studies because we don't have the ability," underscores how poverty restricts opportunities for educational advancement. This lack of access perpetuates cycles of poverty, limiting the community's ability to break free from socio-economic marginalization.

To address these challenges, initiatives such as establishing local schools, providing scholarships, and improving transportation infrastructure are critical.

Moreover, integrating culturally relevant curricula can foster greater engagement and respect for Indigenous knowledge systems.

Gender Disparity Gender disparities in education and decision-making are prominent in the Monda community, reflecting broader societal inequities. Interestingly, Kanika Munda observes that "girls are more advanced than boys" in her community, highlighting the determination and resilience of young women to overcome systemic barriers. Despite this, women face significant obstacles in accessing leadership roles and participating in decision-making processes. Ruma Munda's observation, "Women are excluded from decisions unless they are working or earning," illustrates how economic contributions are often a prerequisite for women's inclusion in community matters. This exclusion perpetuates gender-based inequalities and limits the potential for collective progress. Sonika Munda's statement, "If a girl passes SSC, her parents rush to get her married," highlights the societal pressures that prioritize marriage over education for girls. This practice not only curtails their educational aspirations but also reinforces patriarchal norms that limit women's autonomy. Promoting gender equity requires community-driven interventions that challenge discriminatory norms and practices. Empowering women through education, creating safe spaces for dialogue, and fostering mentorship opportunities can help dismantle these barriers.

Early Marriage Early marriage remains a pervasive issue in the Monda community, deeply rooted in poverty and traditional norms. Kanika Munda explains, "Parents pressure their daughters to marry early because of poverty." Economic insecurity drives families to view early marriage as a means of reducing financial burdens, often at the expense of girls' health and education. The personal experiences of youth further illustrate the profound impact of this practice. Ruma Munda shares, "I was married off at 13. My mother opposed it, but my father wouldn't listen." This narrative underscores the lack of agency afforded to young girls and their families in making life-altering decisions. Early marriage not only disrupts education but also exposes girls to health risks, as noted by Sonika Munda: "Girls should not get married early as it leads to malnutrition and health issues." Addressing early marriage requires a multifaceted approach. Economic empowerment programs for families, community education on the impacts of early marriage, and enforcement of legal protections are essential steps. Additionally, fostering intergenerational dialogue can help shift cultural attitudes and prioritize girls' education and well-being. The challenges of education and gender in the Monda Indigenous community are deeply intertwined with systemic inequities and socio-economic constraints. The youth's narratives highlight the urgency of addressing access to education, gender disparities, and early marriage to create a more equitable and just society. By investing in education, challenging discriminatory norms, and empowering women and girls, the community can build a future where every individual has the opportunity to thrive.

The experiences of Monda Indigenous youth reveal both the barriers and opportunities for advancing education and gender equity. Their resilience and determination serve as a foundation for transformative change. Moving forward, integrating their perspectives into policy and practice can ensure that interventions are inclusive and culturally relevant. By fostering collaboration and prioritizing the voices of those most affected, we can create pathways for sustainable development and social justice.

Livelihood and Poverty Livelihood and poverty are deeply entrenched issues for the Monda Indigenous community, shaping their day-to-day realities and long-term prospects. For the Monda youth, these challenges manifest as limited job opportunities, persistent economic struggles, and seasonal dependency that exacerbate poverty and marginalization. These lived experiences provide a vivid understanding of how systemic barriers perpetuate economic inequities, while also highlighting the resilience and adaptive strategies employed by the community. This section delves into the subthemes of limited job opportunities, economic struggles, and seasonal dependency, shedding light on the interconnectedness of livelihood and poverty in the Monda Indigenous context. The relationship between livelihood and poverty is complex and multifaceted, particularly for the Monda Indigenous community. Their narratives reveal not only the structural challenges they face but also the innovative ways in which they navigate these difficulties to sustain themselves and their families.

Limited Job Opportunities Limited access to stable and diverse job opportunities is a recurring challenge in the Monda community. As Kanika Munda states, "People here cannot study due to poverty. Boys go to brickfields to work." The necessity for boys to forgo education and enter the labor force at a young age perpetuates cycles of poverty and limits future opportunities for economic mobility. Brickfields are one of the few sources of employment available to Monda families. Ruma Munda shares, "Male family members go to work in brickfields for six months in a year." This seasonal migration for work not only disrupts family structures but also creates a dependency on a single, often exploitative, industry.

Sonika Munda's observation, "There's no work here; people depend on the Sundarbans," highlights the scarcity of local employment opportunities. Dependence on natural resources like the Sundarbans for livelihood makes the community vulnerable to environmental changes and overexploitation. The lack of job diversification exacerbates poverty, leaving families with few options to improve their circumstances. Efforts to address these issues should focus on creating sustainable and locally relevant job opportunities. Training programs, cooperative businesses, and investments in community-led enterprises can help diversify income sources and reduce dependence on exploitative industries.

Economic Struggles Economic struggles are a defining aspect of life in the Monda community. Kanika Munda notes, "We live off savings during disasters," underscoring the precarious financial situations that force families to deplete their limited resources during crises. The lack of financial resilience makes recovery from disasters an ongoing challenge.

Ruma Munda's statement, "We don't have an alternative earning source; loans are the only option," highlights the community's reliance on borrowing to meet basic needs. This dependence on loans often leads to cycles of debt that further entrench poverty.

Nutritional deprivation is another critical issue stemming from economic struggles. Ruma Munda shares, "We eat only lentils and potatoes; we can't afford fish or good food." The inability to access a diverse and nutritious diet has long-term implications for the community's health and well-being. To alleviate economic struggles, initiatives such as microcredit programs, community savings groups, and access to affordable, nutritious food are vital. Additionally, integrating disaster preparedness and financial literacy programs can help families build resilience against future economic shocks.

Seasonal Dependency Seasonal dependency on specific economic activities significantly impacts the livelihoods of the Monda community. As Ruma Munda explains, "During monsoon, we plant rice, and in summer, we plant IRRI rice. The rest of the time, we sit idle." This reliance on seasonal agriculture creates periods of inactivity that limit consistent income generation.

The community's dependency on seasonal migration for brickfield work further exacerbates economic instability. Ruma Munda states, "When men return from brickfields, families can eat better, but it's temporary." This temporary reprieve underscores the unsustainable nature of such livelihoods and the ongoing struggle for economic stability. The toll of seasonal migration extends beyond finances, affecting social cohesion and family dynamics. Sonika Munda notes, "Brickfield jobs keep men away from families for half the year." The absence of male family members disrupts caregiving roles and places additional burdens on women and children. Addressing seasonal dependency requires a focus on diversifying livelihood options and promoting year-round economic activities. Agricultural innovation, sustainable resource management, and skills training can empower the community to break free from the constraints of seasonal dependency. Livelihood and poverty are deeply interconnected issues that shape the experiences of the Monda Indigenous community. The youth's narratives highlight the systemic barriers and structural inequities that perpetuate economic struggles while also showcasing their resilience and adaptability. Addressing these challenges requires a holistic approach that combines economic empowerment, sustainable development, and culturally relevant interventions.

The experiences of Monda Indigenous youth provide critical insights into the challenges and opportunities for improving livelihoods and reducing poverty. By centering their voices in policy-making and program design, it is possible to create solutions that are both sustainable and inclusive. Collaborative efforts that prioritize community-driven development and capacity-building can pave the way for a future where economic stability and equity are within reach for all members of the Monda community.

Cultural Identity and Resilience Cultural identity and resilience serve as the cornerstone of the Monda Indigenous community, providing a sense of belonging and strength amidst adversity. For Monda youth, their cultural practices, community challenges, and aspirations for change are deeply intertwined with their lived realities and dreams for the future. This theme explores how Indigenous practices sustain cultural identity, the challenges faced by the community in preserving these traditions, and the transformative aspirations of Monda youth to uplift their community. By centering the voices of Monda youth, we gain valuable insights into the enduring spirit and resilience of their cultural identity. The narratives of Monda youth illuminate the profound role of cultural identity in fostering resilience. Through their stories, we see how Indigenous practices, community challenges, and aspirations for change collectively shape their experiences and outlook.

Indigenous Practices The Monda community's Indigenous practices are deeply rooted in their reverence for nature and spirituality. Ruma Munda's statement, "We worship Ganga, Lord Shiva, and Basumoti, who represent water, air, and earth," reflects the community's intrinsic connection to natural elements. These practices are not only spiritual but also a means of fostering environmental stewardship and collective well-being. Purification rituals, as described by Ruma Munda, "We sprinkle water on our heads before pujas to purify ourselves," symbolize the community's emphasis on purity and respect for sacred traditions. These rituals connect individuals to their heritage and instill a sense of cultural pride. Festivals like Durga Puja serve as communal events that reinforce bonds and collective identity. As Ruma Munda shares, "We celebrate Durga Puja and contribute together because it's costly." This act of collective contribution demonstrates the community's resilience and commitment to preserving their cultural practices despite economic hardships. Indigenous practices are vital in maintaining cultural identity and fostering resilience. Efforts to document and revitalize these traditions, along with promoting cultural education, can ensure their continuation for future generations.

Community Challenges The preservation of cultural identity is not without challenges. Ruma Munda's statement, "Mainstream communities mock our religion, saying disrespectful things," highlights the discrimination and marginalization faced by the Monda community. Such mockery not only undermines their traditions but also creates barriers to self-expression and dignity. Fear of attacks during festivals further underscores the precariousness of cultural preservation. Ruma Munda's observation, "We are afraid of attacks during festivals; idols are expensive and could be destroyed," reflects the vulnerability of the community in practicing their faith openly. This fear diminishes the joy and unity that these celebrations are meant to foster.

The perception of backwardness within the community, as expressed by Sonika Munda, "The Munda community is small and backward, which makes me sad," reveals the internalized effects of systemic marginalization. This sentiment underscores the need for initiatives that empower the community and challenge stereotypes.

Addressing these challenges requires creating safe spaces for cultural expression, combating discrimination, and promoting the contributions of the Monda community. Advocacy and inclusion efforts can help bridge the gap between mainstream society and Indigenous communities.

Aspirations for Change Despite the challenges they face, Monda youth possess a profound sense of hope and determination to uplift their community. Kanika Munda's dream, "I dream of studying, helping people, and developing the Munda community," embodies the aspirations of a generation determined to create positive change.

Ruma Munda's vision of leadership, "If I were a leader, I would prevent teenagers from going to brickfields," reflects a desire to protect and empower the younger generation. This perspective highlights the importance of education and alternative livelihood opportunities in breaking cycles of poverty. The commitment to improving basic living conditions is evident in Sonika Munda's statement, "I would set up a school and ensure fresh water for my community." Such aspirations emphasize the importance of addressing fundamental needs as a pathway to community development and resilience. Supporting the aspirations of Monda youth requires investment in education, leadership development, and community-driven projects. By nurturing their potential, we can create pathways for sustainable and inclusive growth. Cultural identity and resilience are integral to the Monda Indigenous community's survival and growth. The stories of Monda youth reveal the richness of their cultural practices, the challenges they face in preserving their heritage, and their unwavering aspirations for a better future. Addressing these challenges and supporting their aspirations are essential steps toward fostering a thriving and resilient community.

The perspectives of Monda youth provide a powerful testament to the enduring spirit of their cultural identity. By amplifying their voices and investing in culturally grounded initiatives, we can build a future where their heritage is celebrated, and their aspirations are realized. Collaborative efforts that prioritize cultural preservation, community empowerment, and youth leadership will pave the way for a resilient and inclusive society.

Discussion: The Critical Role of Monda Indigenous Youth Perspectives in Climate Crisis Adaptation and Self-Determination

The perspectives of Monda Indigenous youth provide invaluable insights into the multifaceted impacts of the climate crisis and the adaptive strategies required to address these challenges. Rooted in a deep connection to their land and cultural traditions, these youth embody resilience and innovation in the face of environmental, social, and economic adversities. Their lived experiences and voices are critical for understanding the climate crisis and fostering solutions that are not only sustainable but also equitable and culturally relevant. This discussion explores why centering Monda Indigenous youth perspectives is vital for effective climate adaptation and highlights the importance of youth self-determination as a pathway to transformative change.

Monda Youth Perspectives: Bridging Knowledge Systems

Indigenous knowledge systems have long been recognized as vital for understanding and addressing environmental challenges. Monda youth, who are deeply rooted in these traditions, offer unique perspectives that bridge traditional ecological knowledge and contemporary scientific approaches (Berkes, 2009; Nakashima et al., 2012). For instance, their understanding of salinity management and rainwater harvesting reflects centuries of lived experience intertwined with innovative problem-solving (Ford & Pearce, 2010; Pretty & Smith, 2004). These perspectives not only enrich climate adaptation strategies but also challenge the dominance of top-down approaches that often marginalize Indigenous voices. Traditional ecological knowledge (TEK) informs practices such as mangrove restoration, which serves as a natural buffer against cyclones and tidal surges. The integration of TEK with modern climate science can lead to hybrid solutions that are both effective and culturally appropriate. For example, salinity-resistant agricultural techniques, as emphasized by Monda youth, demonstrate the potential for co-creating knowledge systems that respect Indigenous traditions while addressing contemporary challenges (Hauser, 2018; Kimmerer, 2013).

Youth Perspectives on Environmental Transformations The Monda youth are acutely aware of the environmental transformations disrupting their communities. These include rising sea levels, salinity intrusion, cyclones, and flooding, which significantly impact agriculture, biodiversity, and access to freshwater (Berkes, 2009; Nakashima et al., 2012). As articulated by Kanika Munda, high tides, cyclones, and river erosion are recurring threats that destabilize their ecological balance and livelihood security. Such environmental challenges exacerbate existing vulnerabilities, leaving the community with few resources to adapt (Adger, 2000; Walker et al., 2004).

Youth Perspectives on Salinity and Seasonal Dependency Salinity intrusion poses an additional layer of difficulty, affecting agricultural productivity and access to clean water. Kumila Munda highlights this struggle, noting that "saline water makes it difficult to live." This intrusion disrupts not only drinking water supplies but also agricultural systems that are critical for food security, thereby threatening both the physical and cultural survival of the community (Ford & Pearce, 2010; Pretty & Smith, 2004). Economic instability further compounds the struggles faced by the Monda youth. Shrimp farming, often touted as a profitable venture, has led to a loss of agricultural diversity and increased soil salinity. As noted by Sonika Munda, "Rich people are taking land on lease and cultivating shrimp," a practice that marginalizes small-scale farmers and entrenches economic disparities (Escobar, 1995; Hauser, 2018). Such inequities force many families into cycles of poverty, limiting their ability to invest in education or health care (Hall & Patrinos, 2006). The challenges of seasonal dependency also underline the precariousness of the community's livelihoods. Seasonal migration to brickfields leaves families divided for much of the year, placing additional burdens on women and children. This dependency on exploitative industries highlights the urgent need for diversified and sustainable economic opportunities (Leichenko & O'Brien, 2008; Wolfe, 2006).

Centering Youth in Climate Adaptation Monda youth perspectives are particularly valuable because they reflect the lived realities of a generation inheriting the brunt of climate change impacts. As noted in the findings, environmental changes such as cyclones, salinity intrusion, and water scarcity disproportionately affect their livelihoods, health, and education. By centering their voices, policymakers and researchers can gain a nuanced understanding of these challenges, paving the way for targeted interventions that address root causes rather than symptoms (Leichenko & O'Brien, 2008; Whyte, 2017). Youth-led initiatives are emerging as powerful tools for climate action. For example, Monda youth have emphasized the importance of community-driven efforts such as freshwater conservation and disaster preparedness. These initiatives not only enhance community resilience but also foster a sense of ownership and agency among young people (Dudgeon & Berkes, 2003; Wilson, 2008). By supporting youth leadership, stakeholders can ensure that adaptation strategies are informed by local knowledge and priorities, increasing their effectiveness and sustainability.

The Role of Youth Self-Determination Self-determination is a cornerstone of Indigenous resilience and a fundamental right for Indigenous youth. For Monda youth, self-determination encompasses the ability to shape their futures, protect their cultural heritage, and lead community development efforts. This is particularly important in the context of climate change, where external interventions often fail to address the unique needs and aspirations of Indigenous communities (Coulthard, 2014; Smith, 2012). Self-determination also involves challenging systemic inequities that perpetuate marginalization. Monda youth have highlighted the need for equitable access to resources, education, and decision-making processes. By advocating for policies that prioritize their voices and recognize their contributions, these youth are laying the groundwork for more inclusive and just climate adaptation strategies (Adger, 2000; Escobar, 1995).

Resilience Through Education and Empowerment Education is a critical pathway for empowering Monda youth and fostering resilience. As noted in the findings, geographical and economic barriers often limit access to quality education, particularly for girls. Addressing these barriers is essential for unlocking the potential of

Monda youth as leaders in climate adaptation (Battiste, 2002; Hall & Patrinos, 2006). Integrating Indigenous knowledge into educational curricula can further strengthen youth resilience. By valuing and teaching traditional ecological practices alongside modern science, educational programs can bridge generational divides and equip young people with the tools to address complex challenges (Kimmerer, 2013; Nakashima et al., 2012). Additionally, mentorship programs and leadership training can nurture youth agency, enabling them to advocate for their communities and drive transformative change (Coates, 2004; Whyte, 2017).

Addressing Systemic Inequities Systemic inequities, such as economic marginalization and cultural discrimination, pose significant barriers to the self-determination of Monda youth. Shrimp farming and seasonal migration, for instance, have created cycles of poverty that limit opportunities for education and development (Hauser, 2018; Wolfe, 2006). Tackling these inequities requires a holistic approach that combines economic empowerment with cultural preservation. Community-driven development initiatives, such as cooperative businesses and sustainable agricultural practices, can provide alternative livelihoods that align with Indigenous values and traditions. By investing in such initiatives, stakeholders can reduce dependency on exploitative industries and enhance financial resilience within the Monda community (Leichenko & O'Brien, 2008; Pretty & Smith, 2004).

The Transformative Potential of Cultural Identity Cultural identity is a source of strength and resilience for Monda youth. Their commitment to preserving Indigenous practices, such as worshipping natural elements and celebrating communal festivals, fosters a sense of belonging and solidarity (Berkes, 2009; Kimmerer, 2013). These traditions not only sustain the community's spiritual well-being but also serve as a foundation for environmental stewardship. However, cultural identity is often undermined by systemic discrimination and social marginalization. Monda youth have expressed concerns about mockery of their traditions and the fear of attacks during festivals. Addressing these challenges requires creating safe spaces for cultural expression and promoting intercultural dialogue to combat stereotypes and biases (Coulthard, 2014; Smith, 2012).

Moving Toward Inclusive Climate Action Inclusive climate action necessitates centering the perspectives of those most affected by environmental changes. For the Monda Indigenous youth, this means not only recognizing their resilience but also amplifying their voices in decision-making processes. Collaborative efforts that integrate traditional knowledge, community participation, and youth leadership can pave the way for sustainable and equitable solutions (Walker et al., 2004; Wilson, 2008). Furthermore, fostering intergenerational and intercultural collaborations can enhance the effectiveness of climate adaptation strategies. By bringing together diverse perspectives and expertise, these collaborations can address the complexity of the climate crisis while promoting social cohesion and mutual learning (Bohensky & Maru, 2011; Nakashima et al., 2012).

The perspectives of Monda Indigenous youth are indispensable for addressing the climate crisis and fostering sustainable adaptation strategies. Their lived experiences, rooted in traditional ecological knowledge and resilience, provide a valuable lens through which to understand and tackle complex environmental, social, and economic challenges. By prioritizing youth self-determination and integrating their voices into policy and practice, stakeholders can co-create solutions that are inclusive, equitable, and culturally relevant. Ultimately, the empowerment of Monda youth as leaders and change-makers holds the potential to drive transformative change, not only for their community but also for global efforts to combat the climate crisis.

Chapter 5
Human-Created Flood and Land-Based Adaptation from Marma Indigenous Communities

This chapter explores the serious negative impact of human-created floods on the Marma Indigenous communities in the Chittagong Hill Tracts of Bangladesh, emphasizing the systemic vulnerabilities and resilience strategies developed by these communities. Anthropogenic activities such as deforestation, unregulated dam construction, and climate change exacerbate flooding, disrupting livelihoods and cultural practices. The analysis highlights the socioeconomic challenges faced by the Marma, including loss of property, displacement, health crises, and systemic discrimination in disaster relief efforts. Despite these adversities, the Marma people exemplify resilience through land-based adaptation strategies such as elevated housing structures (\u201Cmachang\u201D), floating agriculture, and collective resource management. Central to these approaches is the preservation of indigenous knowledge, passed through generations, which offers practical and sustainable solutions to flood mitigation. The chapter underscores the need for integrating Marma adaptive strategies into national and global climate policies, advocating for collaborative frameworks that prioritize equity, environmental conservation, and resource transparency. By centering the voices of Indigenous communities, this study not only sheds light on their struggles but also positions their innovations as vital contributions to sustainable and inclusive climate adaptation strategies globally.

Floods, long regarded as natural disasters, are increasingly intensified by human activities, turning them into anthropogenic crises. Activities such as deforestation, unregulated dam construction, urbanization, and the broader impacts of climate change have exacerbated the frequency and severity of flooding. For Indigenous communities like the Marma in the Chittagong Hill Tracts (CHT) of Bangladesh, these human-created floods are not just environmental phenomena but existential threats. The Marma, who have lived harmoniously with their natural environment for centuries, now face unprecedented challenges that disrupt their livelihoods, cultural practices, and survival. Their land-based climate adaptation strategies, deeply rooted in indigenous knowledge, offer a beacon of hope for building resilience against floods. This section delves into the unique challenges posed by

R. Datta, *Decolonizing Climate Adaptation*,
https://doi.org/10.1007/978-981-96-9140-1_5

human-created floods to the Marma communities and underscores the critical importance of their adaptive strategies in the broader context of sustainable climate adaptation.

Human-induced changes to the environment have transformed the nature of floods in the CHT. Deforestation is one of the primary drivers, as the removal of forest cover reduces the soil capacity to absorb rainfall, increasing runoff and the likelihood of downstream flooding (Ahmed, Haque, & Neelormi, 2019). In addition, unregulated dam construction disrupts the natural flow of rivers, creating upstream flooding during heavy rains. As one Marma resident recounted, "There was huge heavy rain in Ruma and Thanchi that has caused this flood" learned from the Marma community conversation).

Climate change has further amplified these risks by altering monsoon patterns and intensifying rainfall events. The Marma people, who depend on seasonal cycles for their agricultural practices, now find these cycles increasingly unpredictable (Chowdhury, 2008). Heavy rainfall events not only overwhelm local water systems but also exacerbate soil erosion, further reducing the land's productivity (Haque et al., 2019).

Socioeconomic Impacts of Floods The Marma communities reliance on agriculture and livestock makes them particularly vulnerable to the impacts of floods. Fields are inundated, crops are destroyed, and livestock is swept away, leaving families without their primary sources of income and sustenance. "Our house got destroyed. Many things got lost with the flood water," shared one resident (Learned from Marma Community conversation). The loss of agricultural land cornerstone of the Marma economy families into cycles of poverty and food insecurity (Rahman et al., 2020). Displacement is another critical issue. Rising floodwaters compel entire families to leave their homes, often seeking refuge on roads, embankments, or temporary shelters that lack basic amenities (Ahmed et al., 2015). "We had no money to rent a home, so had to stay on the road," said a displaced family member, reflecting the financial and emotional toll of such disruptions (Learned from Marma community conversation, 2023). For children, the impacts are particularly severe, as displacement disrupts their education and exposes them to heightened health risks (Khan, 2025).

Health Crises Exacerbated by Floods Floodwaters, often contaminated with waste, lead to the spread of waterborne diseases such as diarrhea, cholera, and skin infections (Haque et al., 2019). Poor sanitation facilities in temporary shelters exacerbate these conditions, creating public health crises in affected areas. "People suffered a lot from the sanitation problem because everything was underwater," noted a community member (learned from Marma Community conversation). Children are especially vulnerable, with pneumonia and other respiratory illnesses becoming common due to prolonged exposure to damp and unsanitary environments (Rahman et al., 2020). The financial burden of healthcare further strains families already struggling to rebuild their lives. "We had to buy medicine for that. We had no money to buy medicines," one resident recounted, highlighting the dire state of health services in the region (Learned from Marma Community conversation).

Systemic Neglect and Discriminatory Practices While environmental factors drive the physical impacts of floods, systemic neglect and inequities compound their effects on the Marma communities. Discriminatory practices in disaster relief often exclude Indigenous groups from receiving adequate support. "Leaders made a list of the flood-affected people, but then cut [erased] those names," shared a frustrated resident (Learned Marma Community conversation). Such practices reveal the deep-rooted biases in resource distribution systems, leaving marginalized communities disproportionately underserved (Chowdhury, 2008). This neglect is further exacerbated by the lack of political representation and limited inclusion of Indigenous perspectives in disaster management frameworks. Without meaningful engagement, the specific needs and vulnerabilities of the Marma communities remain unaddressed, perpetuating cycles of marginalization (Haque et al., 2019).

The Critical Role of Marma Land-Based Climate Adaptation In the face of these challenges, the Marma people have developed innovative land-based adaptation strategies that draw on centuries of indigenous knowledge and environmental stewardship. These practices not only mitigate the immediate impacts of floods but also contribute to long-term resilience and sustainability.

One of the most effective adaptations is the construction of elevated housing structures, locally known as “machang.” Built on stilts using bamboo and other locally sourced materials, these homes are designed to withstand rising floodwaters. “Machang ensures the safety of people and their possessions during floods,” explained a Marma elder (Learned from Marma Community conversation). By integrating traditional architectural techniques with practical considerations, these structures exemplify how indigenous knowledge can inform sustainable housing solutions (Rahman et al., 2020).

Floating agriculture is another adaptation strategy that highlights the Marma community ingenuity. By creating buoyant platforms made from water hyacinths and other materials, they can cultivate crops like vegetables and spices even during prolonged flooding (Haque et al., 2019). “Floating beds have been a lifeline during the floods,” remarked a resident, underscoring their importance for food security and economic stability (Learned from Marma Community conversation).

The Marma strong tradition of collective resource management becomes particularly critical during crises. By pooling resources and labor, they repair damaged infrastructure, construct temporary embankments, and secure clean drinking water. These collaborative efforts strengthen social cohesion and foster a sense of unity that is essential for resilience (Khan, 2025).

Central to the Marma adaptive strategies is the preservation and application of indigenous knowledge. Through oral traditions, they pass down critical information about weather patterns, flood-prone seasons, and survival techniques. “Our ancestors taught us how to prepare for floods and protect our homes,” shared a community elder (Learned from Marma community conversation). This knowledge not only enhances preparedness but also reinforces cultural identity and pride (Rahman et al., 2020).

The Need for Policy Integration. While the Marma adaptation strategies are exemplary, they cannot address the root causes of human-created floods alone. Integrating these practices into national and regional climate policies is essential for creating sustainable solutions. Collaborative efforts between Indigenous communities, governments, and development agencies can foster more inclusive and effective disaster management systems (Ahmed et al., 2015). As studies suggested that the key interventions should include:

- Infrastructure development: Constructing flood-resistant infrastructure such as elevated roads and permanent shelters (Chowdhury, 2008).
- Healthcare access: Establishing mobile health clinics to address health crises during floods (Rahman et al., 2020).
- Environmental conservation: Promoting reforestation and sustainable watershed management to mitigate flood risks (Haque et al., 2019).
- Equitable resource distribution: Ensuring transparency and fairness in aid allocation to eliminate systemic biases (Khan, 2025).

The Marma Indigenous communities'resilience in the face of human-created floods underscores the critical importance of integrating Indigenous knowledge into broader climate adaptation strategies. Their land-based practices, from elevated housing and floating agriculture to community-based resource management, offer powerful lessons for building sustainable and inclusive resilience systems. However, systemic neglect and discriminatory practices continue to hinder their efforts (Table 5.1). Addressing these challenges through inclusive policies and collaborative frameworks is essential for empowering the Marma people and ensuring their survival in a changing climate. By recognizing and valuing the contributions of Indigenous communities like the Marma, we can pave the way for more equitable and effective climate adaptation initiatives. Their ingenuity and resourcefulness provide a blueprint for tackling the complex challenges of human-created floods, highlighting the transformative potential of culturally informed solutions.

Findings

Impact of Floods Floods, a natural disaster exacerbated by human activities, significantly impact vulnerable communities worldwide. For the Marma Indigenous communities, a minority group in the Chittagong Hill Tracts of Bangladesh, the challenges posed by floods are particularly acute. Human-induced environmental changes—such as deforestation, unregulated dam construction, and climate change—have intensified the frequency and severity of flooding (Fig. 5.1). While the Marma people are not responsible for these large-scale ecological disruptions, they are often the most affected, given their geographical and socio-economic vulnerabilities. This essay explores the devastating impact of floods on the Marma community and examines how their traditional knowledge and adaptive strategies serve as a means of survival and resilience. By focusing on the destruction of prop-

Table 5.1 shows themes and subthemes from the monda indigenous youth perspectives on climate crisis and land-based adaptions

Main themes	Subthemes
Human-created floods	– Environmental drivers (deforestation, unregulated dams, climate change)
	– Intensification of seasonal rainfall and unpredictability
Socioeconomic impacts	– Loss of property, crops, and livestock
	– Displacement challenges and shelter inadequacies
	– Economic cycles of poverty and food insecurity
Health impacts	– Spread of waterborne diseases (diarrhea, cholera)
	– Sanitation crises in temporary shelters
	– Vulnerability of children to respiratory illnesses
Systemic inequities	– Discrimination in aid distribution
	– Lack of political representation
Indigenous adaptation strategies	– Elevated housing structures (\u201Cmachang\u201D)
	– Floating agriculture for food security
	– Community-based resource management
	– Preservation of indigenous knowledge
Policy integration	– Collaborative frameworks with governments and NGOs
	– Infrastructure development and healthcare access
	– Reforestation and environmental conservation
	– Equitable and transparent resource distribution

Fig. 5.1 Illustrates the impact of human activities on the Marma Indigenous communities in the Chittagong Hill Tracts (CHT) of Bangladesh, highlighting the severe consequences of the 2023 floods. These floods were primarily caused by the encroachment on rivers, canals, and other natural water flow pathways

erty and livelihoods, displacement challenges, and health impacts, this analysis underscores the need for a holistic approach to flood management and adaptation.

Destruction of Property and Livelihoods Floods have a catastrophic effect on the Marma community's homes, possessions, and means of livelihood. As their villages are often situated in low-lying areas near rivers, they are highly susceptible to flooding, which sweeps away homes, livestock, and crops. A Marma Indigenous Marma Indigenous resident recounts, "Our house got destroyed. Many things got lost with the flood water." These sentiments reflect the widespread devastation experienced by the community. The loss of agricultural land, a primary source of income, exacerbates the economic distress. In many cases, families are left with no external support to rebuild their lives, forcing them to rely on personal initiative and hard work: "They have survived by their own initiatives, by their own money, by their own hard work." The destruction extends beyond the physical loss of property to include livestock, which serves as a critical resource for sustenance and economic stability. "People lost their houses and property, livestock, etc.," shares another community member. With limited financial resources and external aid, the cycle of poverty deepens, leaving many without hope of recovery.

Displacement and Shelter Challenges The rising floodwaters force entire families to leave their homes, often without warning. Many find themselves living on roads, embankments, or in overcrowded shelters. These temporary refuges, while providing safety from the immediate danger, often lack adequate facilities. One family recalls, "We had no money to rent a home, so had to stay on the road." Such conditions are not only physically taxing but also emotionally draining, especially for families with children. The shelters—though overcrowded—offered some semblance of security, particularly for teenagers: "Teenagers didn't face any problems. Everyone was secured during the flood when they were staying in the shelter." Despite the resilience displayed by the Marma community, the lack of long-term shelter solutions often leaves them vulnerable to subsequent floods. Without permanent relocation options or government assistance, many are forced to return to flood-prone areas, perpetuating a cycle of displacement and vulnerability.

Health Impacts Floods have severe health implications, particularly for marginalized communities like the Marma. Contaminated floodwaters lead to the spread of waterborne diseases, while inadequate sanitation facilities exacerbate these problems. "People suffered a lot from the sanitation problem because everything was underwater," says a community member, highlighting the dire conditions faced during floods.

The financial burden of healthcare is another significant issue. Many families struggle to afford medical treatment, as one resident explains: "We have faced some problems and had to buy medicine for that. We had no money to buy medicines." For children, prolonged exposure to floodwaters increases the risk of diseases like pneumonia, further straining the community's limited resources: "Another thing, I noticed pneumonia among children."

Land-Based Adaptation Strategies for Impact Assessment Despite these challenges, the Marma community has demonstrated remarkable resilience by employing traditional knowledge and innovative adaptation strategies. These land-based approaches not only help mitigate the impact of floods but also preserve the cultural heritage of the community.

Elevated Housing Structures To combat the recurrent flooding, the Marma people have developed elevated housing structures, often built on stilts. These traditional homes, known as "machang," are constructed using locally available materials like bamboo and wood. The raised design prevents floodwaters from entering living spaces, ensuring the safety of both people and possessions. By adapting their housing structures to the natural environment, the Marma people exemplify sustainable living practices.

Floating Agriculture Floating agriculture, a practice widely used in flood-prone areas, has been adopted by the Marma community as a means of ensuring food security during and after floods. Using water hyacinths and other buoyant materials, they create floating beds for growing crops like vegetables and spices. This innovative approach not only provides sustenance during floods but also generates income, reducing economic dependency on external aid.

Community-Based Resource Management

The Marma community has a long tradition of collective resource management, which becomes particularly crucial during floods. By pooling resources and labor, they repair damaged infrastructure, build temporary embankments, and secure clean drinking water. These collaborative efforts strengthen social bonds and foster a sense of unity, which is essential for overcoming the challenges posed by natural disasters.

Preservation of Indigenous Knowledge

Indigenous knowledge plays a vital role in the Marma community's adaptation strategies. Through oral traditions, they pass down valuable information about weather patterns, flood-prone seasons, and survival techniques. This knowledge enables them to prepare for floods more effectively, minimizing the loss of life and property.

The Need for External Support While the Marma community's adaptation strategies are commendable, they are not sufficient to address the growing challenges posed by human-induced floods. Government intervention and external support are essential to complement these efforts. Key areas for support include:

- *Infrastructure development*: Building flood-resistant infrastructure, such as elevated roads and permanent shelters, can provide long-term solutions to displacement challenges.
- *Access to healthcare*: Establishing mobile health clinics and ensuring the availability of affordable medicines can significantly improve the community's resilience to health crises during floods.
- *Financial assistance*: Providing financial aid or low-interest loans can help families rebuild their homes and livelihoods after a flood.
- *Environmental conservation*: Addressing the root causes of human-induced floods, such as deforestation and unregulated dam construction, is crucial for long-term sustainability.

The Marma Indigenous communities in Bangladesh exemplify resilience and ingenuity in the face of human-induced floods. Through elevated housing, floating agriculture, and community-based resource management, they have developed effective adaptation strategies that align with their cultural and environmental contexts. However, these efforts alone cannot address the systemic issues underlying flood vulnerability. A collaborative approach—combining indigenous knowledge with government support and environmental conservation—is essential for sustainable flood management. By recognizing and valuing the contributions of Indigenous communities like the Marma, policymakers and development agencies can create more inclusive and effective solutions. In doing so, we can not only mitigate the impact of floods but also honor the resilience and resourcefulness of the Marma people, ensuring their survival and prosperity in a changing world.

Discriminatory Relief and Rehabilitation System Towards Indigenous and Minority Communities Disasters, whether natural or human-induced, often expose and exacerbate the inequalities entrenched within societal structures. For Indigenous and minority communities, discriminatory practices in relief and rehabilitation efforts deepen their vulnerability and marginalization. These inequities manifest through inconsistent government and NGO support, disparities in aid distribution, and eroding community solidarity. This essay delves into the challenges faced by these communities during crises, highlighting the systemic flaws in relief and rehabilitation mechanisms and underscoring the urgent need for equitable and inclusive disaster management.

Government and NGO Support The role of government and non-governmental organizations (NGOs) in providing relief during disasters is pivotal. However, the distribution of aid has often been inconsistent, with some areas receiving prompt assistance while others are neglected. A resident reflects on the sporadic nature of aid: "The municipality gave water twice. And khichuri twice. There was scarcity of food and water." While some individuals received basic necessities such as dry food, water, candles, and molasses from the army, these provisions were insufficient to meet the needs of the entire affected population. As one community member emphasized, "Government must help them. They need shelter and other essential commodities."

This uneven response underscores the need for a more systematic and inclusive approach to disaster relief. Government assistance, though critical, often fails to reach the most vulnerable communities in a timely and adequate manner. NGOs, while supplementing government efforts, also face limitations in their reach and resource allocation.

Disparities in Aid Distribution One of the most glaring issues in disaster relief is the inequitable distribution of aid, particularly for minority and Indigenous communities. Discrimination, whether overt or systemic, leaves these groups disproportionately underserved.

A resident shares, "Leaders made a list of the flood-affected people, but then cut (erased) those names." This practice of excluding certain individuals from relief eligibility reflects deeply ingrained biases within local leadership structures. Minority individuals often receive less assistance than their counterparts, as highlighted by one member of the community: "Minority people got less assistance after the flood." For some, accessing aid depended on personal connections or informal networks rather than a fair and transparent system. "Some people got 4 kg rice, oil (250 ml), and other reliefs. But we did not get any," reveals another affected individual. Such disparities underscore the systemic flaws in relief distribution, where favoritism and discrimination often leave the most marginalized communities behind.

Community Support and Gaps While communal bonds can serve as a safety net during crises, these networks are not immune to the pressures of poverty and external exploitation. In some areas, community solidarity was evident, with people coming together to support one another during the immediate aftermath of a disaster. However, this unity often wanes over time, leaving many to fend for themselves. "In the community level, people are helpless. They have been living from hand to mouth," shares a resident, highlighting the precariousness of their situation. The strain of poverty weakens traditional support systems, eroding the collective resilience that once defined these communities. The shrinking size and cohesion of Indigenous and minority communities also contribute to the loss of solidarity. "Previously, the community was small. Everyone helped each other," laments a community member, pointing to the challenges of maintaining communal ties in the face of displacement, resource scarcity, and socio-economic pressures.

The Need for Equity and Inclusion Addressing the disparities in disaster relief and rehabilitation requires a multi-faceted approach that prioritizes equity and inclusivity. Key measures include:

Transparent and fair aid distribution. Establishing transparent mechanisms for aid distribution can help mitigate biases and ensure that resources reach the most vulnerable. Independent monitoring bodies and community representatives should be involved in the allocation process to promote accountability.

Strengthening community resilience. Investing in community-based disaster preparedness and response systems can enhance resilience. Training programs, resource-sharing initiatives, and capacity-building efforts can empower communities to better cope with future crises.

Policy reforms. Governments must implement policies that address systemic discrimination in disaster management. This includes mandating the inclusion of minority and Indigenous communities in planning and decision-making processes, as well as ensuring equitable resource allocation.

Support for traditional practices. Recognizing and supporting traditional coping mechanisms and communal support systems can strengthen resilience. By integrating Indigenous knowledge into formal disaster management strategies, policymakers can create culturally sensitive and effective solutions.

Discriminatory relief and rehabilitation systems disproportionately affect Indigenous and minority communities, compounding their struggles during disasters. The inconsistencies in government and NGO support, coupled with systemic biases in aid distribution, highlight the urgent need for reform. Strengthening community resilience, ensuring transparency in resource allocation, and incorporating traditional knowledge into disaster management are critical steps toward creating a more equitable system. By addressing these disparities, we can not only alleviate the immediate suffering of marginalized groups but also foster a more inclusive and just society. As we strive for a future where no one is left behind, it is imperative to center the voices and needs of those who have been historically marginalized, ensuring that relief and rehabilitation efforts are equitable, transparent, and culturally respectful.

Environmental and Social Changes Environmental and social transformations are reshaping the lives of communities worldwide, with Indigenous and minority groups often bearing the brunt of these changes. For communities in Bangladesh, climate change, rapid urbanization, and cultural shifts are driving unprecedented disruptions. These forces not only exacerbate environmental vulnerabilities but also challenge the social fabric and traditional ways of life. This essay explores the intertwined impacts of climate change, population growth, urbanization, and cultural shifts, examining how they contribute to both environmental degradation and social tensions.

Climate Change and Unpredictability Climate change is altering seasonal patterns in Bangladesh, leading to unpredictable weather and extreme events. As one resident observes, "Bangladesh has six seasons. But nowadays we have only two seasons. I define climate change by this change." This sentiment reflects a broader phenomenon of shifting climatic conditions that disrupt traditional agricultural cycles and livelihoods. Heavy rainfall, particularly in upstream regions, has become more frequent, exacerbating flooding and soil erosion. "There was huge heavy rain in Ruma and Thanchi that has caused this flood," notes a community member. These changes underscore the urgent need for adaptive measures to mitigate the impact of climate-induced disasters.

Population Growth and Uneven-Urbanization Rapid population growth and unplanned urbanization are major contributors to environmental degradation. With increasing human settlements encroaching on natural ecosystems, the pressure on rivers, forests, and arable land has intensified. "The river has been captured because of new establishments and population influx," remarks a community member, highlighting the encroachment on vital waterways. Another resident reflects, "In the past, there were no people even before 40 years. But now, there is no place without people." These observations reveal the strain that unchecked development places on the environment, threatening the sustainability of local ecosystems.

The rapid pace of urbanization also disrupts community cohesion, as traditional values and practices are often replaced by modern, individualistic lifestyles. This transition erodes the communal bonds that have historically supported resilience in the face of challenges.

Religious and Cultural Shifts Floods and other natural disasters are sometimes interpreted through a religious or moral lens, reflecting cultural shifts and generational changes. "People became sinners, they don't obey religions. That's why floods happen," explains one community member, showcasing how spiritual beliefs influence perceptions of environmental events. Additionally, changes in cultural harmony are evident. "The harmony is not there, that it used to be in the past," laments another resident. These shifts highlight the complex interplay between environmental changes and evolving social norms, suggesting a need for renewed efforts to foster cultural resilience.

Education and Future of Youth Natural disasters, such as floods, have a profound impact on the education and future of youth, particularly in marginalized communities. The destruction of educational resources and the disruption of schooling create long-term challenges that affect children's prospects and the overall development of their communities. This essay explores the consequences of disasters on education and highlights the critical roles of government and community in recovery efforts.

Loss of Educational Resources Floods often devastate schools and educational materials, leaving students unable to continue their studies. As one community member recounts, "All their educational instruments like books, notebooks got washed away." Such losses disproportionately affect rural and underprivileged students, many of whom already face barriers to accessing quality education.

Students who rely on hostels for their education are particularly vulnerable. "Many students come from afar, stay in hostels and study. They've been affected a lot," shares another resident. Without adequate resources to replace what was lost, many students face prolonged interruptions in their education.

Disruption of Schooling Floods not only destroy resources but also disrupt the daily routines of students, forcing schools to close temporarily. "They couldn't go to school," says a community member, emphasizing the immediate impact on children's learning. While communities often come together during the crisis—"During the flood, all people eat together, there was no problem"—the aftermath presents significant challenges, as families struggle to rebuild their lives and prioritize education amidst competing needs.

Government and Community Roles in Recovery Rebuilding the educational futures of affected youth requires coordinated efforts from both government and local communities. Providing shelter, replacing lost educational materials, and ensuring the continuity of schooling are essential steps in this process. "Government must help them. They need shelter and other essential commodities," urges a community member. Beyond material support, initiatives such as mobile schools, temporary classrooms, and mental health support for students can help mitigate the long-term impact of disasters. Community involvement also plays a crucial role. As one individual recounts, "We used to go to visit their house after the duty of our clinic," illustrating how collective efforts can support recovery. By fostering a sense of solidarity and shared responsibility, communities can create a supportive environment for youth to thrive despite the challenges posed by disasters.

Environmental and social changes, coupled with natural disasters, are reshaping the lives and prospects of communities in Bangladesh. Climate change, population growth, and cultural shifts exacerbate vulnerabilities, while floods disrupt education and threaten the future of youth. Addressing these challenges requires a comprehensive approach that integrates environmental sustainability, equitable resource distribution, and community-led initiatives. By investing in adaptive strategies, fostering cultural resilience, and prioritizing education, we can empower communities to navigate the complexities of a changing world. Through collaboration and a commitment to equity, it is possible to build a more resilient and inclusive future for all.

Minority and Marginalized Communities Disasters often lay bare the systemic inequalities ingrained within societies, disproportionately affecting minority and marginalized communities. For these groups, the impacts of natural calamities like floods are compounded by systemic barriers, neglect, and discrimination. Despite these challenges, such communities exhibit remarkable resilience, often relying on self-reliance and communal support to navigate the aftermath. This essay explores the systemic inequalities faced by minorities, their strategies for survival amidst neglect, and the dynamics of community cohesion and conflict during crises.

Systemic Inequality Minority communities are frequently marginalized in disaster response efforts, with systemic discrimination exacerbating their vulnerability. Reports from affected individuals highlight the absence of financial support and the mismanagement of relief resources: "Minorities have not got any financial assistance that time," one resident shared. Another observed, "Leaders are taking everything in their pockets," pointing to corruption and favoritism in the distribution of aid. This neglect leaves minority groups disproportionately exposed to the impacts of disasters, as they often lack the political representation and social capital necessary to advocate for their needs. The systemic exclusion of these communities from relief efforts underscores the urgent need for equitable disaster management policies that prioritize marginalized populations.

Self-Reliance Amidst Neglect In the face of systemic neglect, minority communities often turn to their own resilience and resourcefulness to survive. "Hindus survived doing by themselves," reflects one resident, illustrating how marginalized groups must rely on their limited means to recover from disaster-induced losses. The scarcity of resources forces many to adopt drastic measures to stretch what little they have. "They have lived by the amount of food for three days that they were supposed to eat for three meals," notes another individual. Such accounts reveal the immense sacrifices made by minority communities to sustain themselves during and after crises. While self-reliance is a testament to the strength of these communities, it also underscores the failure of broader systems to provide adequate support. This gap highlights the critical role of inclusive policies and practices in ensuring that no community is left to fend for itself during disasters.

Community Conflicts and Cohesion Disasters often create temporary moments of unity, as communities come together to survive immediate threats. During the floods, individuals reported instances of communal harmony: "No one thought about anyone's religion, financial status, and social status." These moments of solidarity demonstrate the potential for collective resilience in the face of adversity. However, such cohesion is often fleeting. Longstanding social divides and discrimination quickly resurface once the immediate crisis subsides. "The sincerity they show about us is just a formality," remarks one resident, reflecting the persistent marginalization of minority groups even in times of apparent unity. The re-emergence of these divides underscores the need for sustained efforts to build trust and foster genuine inclusivity within communities. By addressing the root causes of social tensions, it is possible to create a more cohesive and equitable society that can better withstand future challenges.

The experiences of minority and marginalized communities during floods reveal the systemic inequalities that shape disaster response and recovery. While these communities demonstrate remarkable resilience and resourcefulness, their struggles highlight the failures of existing systems to provide equitable support. Addressing these issues requires comprehensive reforms that prioritize inclusion, transparency, and accountability in disaster management. By fostering community cohesion and addressing systemic discrimination, societies can build a more just and resilient future for all. The stories of self-reliance and temporary unity among minority groups serve as a reminder of the potential for collective strength when equity and inclusivity are placed at the forefront of disaster response efforts.

Discussion

Floods have always been a part of the natural environment, but their frequency and severity have increased significantly due to human activities. The Marma Indigenous communities in the Chittagong Hill Tracts (CHT) of Bangladesh are particularly vulnerable to these anthropogenic changes, with their lives, culture, and livelihoods deeply affected. Human-induced factors such as deforestation, unregulated dam construction, and climate change have exacerbated the flooding issue. Deforestation in the CHT has led to soil erosion and reduced the land's capacity to absorb water, causing higher runoff levels during heavy rains. As Ahmed et al. (2015) note, the removal of forest cover not only destabilizes ecosystems but also intensifies the impact of seasonal monsoons. The unregulated construction of dams further disrupts the natural flow of rivers, often creating upstream flooding. A Marma resident vividly described the devastating impact: "There was huge heavy rain in Ruma and Thanchi that has caused this flood" (Learned from Marma Community conversation). Additionally, climate change has altered weather patterns, making rainfall more erratic and intense. According to Chowdhury (2008), these changes have severely disrupted the Marma people agricultural cycles, leading to widespread crop loss and food insecurity. This unpredictability is especially challenging for a community whose economy is primarily reliant on subsistence farming (Rahman et al., 2020).

Socioeconomic Consequences The destruction caused by human-created floods has profound socioeconomic implications for the Marma communities. Floodwaters often destroy homes, crops, and livestock, pushing families into cycles of poverty. A resident lamented, "Our house got destroyed. Many things got lost with the flood water," highlighting the magnitude of the losses experienced (Learned from Marma Community conversation). Livelihood disruptions are compounded by the lack of external financial assistance, forcing families to deplete their limited resources to rebuild (Haque et al., 2019).

Displacement Displacement is another critical issue. Floods force families to seek refuge on roads, embankments, or overcrowded shelters that lack adequate facilities. "We had no money to rent a home, so had to stay on the road," explained one displaced resident, reflecting the economic and emotional toll of these disruptions (Learned from Marma Community conversation). For children, the effects are particularly severe, as displacement disrupts their education and exposes them to heightened health risks (Khan, 2025).

Health Challenges Flood-induced health crises are common among the Marma. Contaminated floodwaters lead to outbreaks of waterborne diseases such as diarrhea, cholera, and skin infections (Haque et al., 2019). Poor sanitation facilities in temporary shelters exacerbate these issues, creating public health emergencies. As one community member noted, "People suffered a lot from the sanitation problem because everything was underwater" (Learned from Marma Community conversation). Children are especially vulnerable to respiratory illnesses such as pneumonia, which are exacerbated by prolonged exposure to damp and unsanitary environments (Rahman et al., 2020). The financial burden of healthcare adds another layer of strain, with families struggling to afford treatment for preventable illnesses. “We had to buy medicine for that. We had no money to buy medicines,” shared a resident, underscoring the critical need for accessible healthcare services in flood-prone areas (Learned from Marma Community conversation).

Systemic Neglect and Discrimination The Marma communities not only face environmental challenges but also systemic neglect and discrimination in disaster relief efforts. Discriminatory practices often exclude Indigenous groups from receiving adequate support. For instance, one resident noted, “Leaders made a list of the flood-affected people, but then cut [erased] those names,” highlighting the inequities in resource distribution (Learned from Marma Community conversation). Without political representation or meaningful inclusion in disaster management frameworks, the specific needs of the Marma are overlooked, perpetuating cycles of marginalization (Chowdhury, 2008).

The Role and Importance of Indigenous Land-Based Adaptation Despite the systemic barriers they face, the Marma communities have developed innovative and sustainable land-based adaptation strategies. These practices, rooted in traditional knowledge, not only mitigate the impacts of floods but also contribute to long-term resilience.

Elevated Housing Structures The Marma construction of elevated housing structures, or "machang," demonstrates their ingenuity in flood adaptation. Built on stilts using bamboo and other locally available materials, these homes protect against floodwaters while preserving cultural heritage. "Machang ensures the safety of people and their possessions during floods," explained a community elder (Learned from Marma Community conversation). This architectural practice aligns with sustainable housing principles and offers a blueprint for flood-resilient construction in other regions (Rahman et al., 2020).

Floating Agriculture Floating agriculture is another innovative adaptation strategy employed by the Marma. By creating buoyant platforms from water hyacinths and other materials, they cultivate crops like vegetables and spices even during prolonged flooding. This method ensures food security and reduces dependency on external aid. “Floating beds have been a lifeline during the floods,” remarked a resident, underscoring the importance of this practice (Learned from Marma Community conversation).

Community-Based Resource Management The Marma tradition of collective resource management is central to their resilience. By pooling resources and labor, they repair damaged infrastructure, construct temporary embankments, and secure clean drinking water. These collaborative efforts strengthen social cohesion and provide a sense of unity during crises (Khan, 2025).

Preservation of Indigenous Knowledge Indigenous knowledge is a cornerstone of the Marma adaptation strategies. Through oral traditions, they pass down valuable insights about weather patterns, flood-prone seasons, and survival techniques. "Our ancestors taught us how to prepare for floods and protect our homes," shared a community elder, emphasizing the importance of intergenerational knowledge transfer.

Pathways for Integration and Policy Support While the Marma adaptation strategies are effective, they require external support to address the root causes of human-created floods. Collaborative efforts between Indigenous communities, governments, and development agencies are essential for fostering inclusive and sustainable disaster management systems. Like other studies, this study also suggested that the key areas for intervention include-

- Infrastructure development: Constructing flood-resistant infrastructure such as elevated roads and permanent shelters to reduce displacement risks (Chowdhury, 2008).
- Healthcare access: Establishing mobile health clinics and ensuring affordable medicines to address health crises during floods (Rahman et al., 2020).
- Environmental conservation: Promoting reforestation and sustainable watershed management to mitigate flood risks (Haque et al., 2019).
- Equitable resource distribution: Ensuring transparency and fairness in aid allocation to eliminate systemic biases (Khan, 2025).

The Marma Indigenous communities resilience in the face of human-created floods highlights the transformative potential of culturally informed adaptation strategies. By integrating traditional knowledge with modern policy frameworks, these practices can serve as models for sustainable climate adaptation worldwide. However, systemic neglect and discriminatory practices continue to undermine their efforts. Addressing these challenges through inclusive policies and collaborative frameworks is critical for empowering the Marma people and ensuring their survival in a rapidly changing climate.

Chapter 6
Resilience, Spirituality, and Environmental Awareness in the Face of Disasters: Perspectives from Marma Indigenous Communities

This chapter examines the resilience, spirituality, and environmental awareness of the Marma indigenous communities in the Chittagong Hill Tracts, Bangladesh, amidst socio-environmental challenges and disasters. It explores the interplay between their spiritual ethos, traditional ecological knowledge, and adaptive strategies that emphasize harmony with nature and community solidarity. Key findings highlight the importance of land-based perspectives, sustainable practices, and cultural preservation in fostering resilience. The chapter underscores the role of government, institutions, and inclusive policies in addressing displacement, corruption, and resource mismanagement. By integrating indigenous knowledge into disaster risk reduction and environmental management, this study offers insights into building equitable and sustainable futures in vulnerable regions.

The Marma indigenous communities of the Chittagong Hill Tracts (CHT) in Bangladesh have long exemplified resilience in the face of socio-economic and environmental adversities. Rooted in a deep connection with nature and a rich cultural and spiritual heritage, the Marma people navigate the challenges posed by environmental degradation, natural disasters, and displacement with remarkable fortitude. This introduction explores the interplay between resilience, spirituality, and environmental awareness within these communities, offering insights into their adaptive strategies and perspectives on sustainability.

Resilience, as a concept, is pivotal to understanding how communities cope with and recover from adversities. The Marma people's resilience is interwoven with their cultural identity, social networks, and traditional knowledge systems (Berkes et al., 2000). For indigenous communities, resilience extends beyond physical survival to encompass the preservation of cultural practices and social cohesion (Adger, 2000). The Marmas' ability to withstand environmental challenges is deeply rooted in their spiritual worldview, which emphasizes harmony with nature and reverence for its resources (Nakashima et al., 2000).

Spirituality plays a central role in Marma culture, guiding their interactions with the environment and shaping their responses to disasters. For the Marmas,

R. Datta, *Decolonizing Climate Adaptation*,
https://doi.org/10.1007/978-981-96-9140-1_6

spirituality is not confined to religious rituals but is embedded in daily life and community practices (Sopher, 1968). This spiritual ethos fosters a profound sense of responsibility towards the environment, as reflected in their traditional land-based practices and beliefs (Chakma, 2016). For instance, the Marma people's reverence for sacred groves and trees underscores their commitment to conserving biodiversity and respecting natural resources (Hossain & Chakma, 2018).

Environmental awareness is another critical dimension of the Marma communities' resilience. Living in a region prone to landslides, floods, and cyclones, the Marmas have developed adaptive strategies that combine traditional ecological knowledge with modern practices (Islam & Walkerden, 2014). Their experiences underscore the importance of integrating Indigenous perspectives into disaster risk reduction and environmental management (Rahman et al., 2012). The Marma people's holistic understanding of their environment, informed by generations of lived experience, offers valuable lessons for sustainable development and climate resilience (Dumont et al., 2017).

Disasters, whether natural or anthropogenic, pose significant challenges to the Marmas' way of life. Climate change exacerbates the frequency and intensity of disasters in the CHT, amplifying vulnerabilities and disrupting traditional livelihoods (Ahmed, Mallick, & Rahman, 2019). In response, the Marma communities rely on their collective strength and cultural practices to adapt and rebuild (Rahman, 2020). For example, communal efforts to restore damaged infrastructure and share resources during crises highlight the importance of social capital in enhancing resilience (Adger et al., 2005).

This study seeks to illuminate the unique perspectives of the Marma indigenous communities on resilience, spirituality, and environmental awareness in the context of disasters. Drawing on qualitative data from ethnographic studies and interviews, it explores how these interrelated dimensions shape the Marmas' adaptive strategies and their vision for sustainable living. By highlighting the Marmas' experiences, this research aims to contribute to a broader understanding of indigenous resilience and its implications for disaster management and environmental sustainability.

The resilience of the Marma communities is not an isolated phenomenon but part of a broader pattern observed among indigenous peoples worldwide. Indigenous communities often possess a wealth of traditional knowledge and practices that enable them to adapt to environmental changes and mitigate disaster risks (Berkes & Folke, 1998). For the Marmas, this resilience is manifested in their land-use practices, water management systems, and community-based disaster preparedness initiatives (Huq et al., 2013). These practices underscore the importance of preserving indigenous knowledge as a valuable resource for addressing contemporary environmental challenges (Nakashima & Roué, 2002).

Spirituality, as a lens for understanding the Marma communities' relationship with nature, offers profound insights into their environmental ethics and values. The Marmas' spiritual practices often emphasize interdependence and reciprocity, fostering a sense of stewardship towards the environment (Sopher, 1968). For instance, rituals and ceremonies associated with planting and harvesting reflect their gratitude for nature's bounty and their commitment to sustainable resource use (Chakma &

Chakma, 2021). These practices highlight the potential of integrating spiritual values into contemporary environmental policies and education (Hossain & Chakma, 2018).

Environmental awareness among the Marmas is shaped by their lived experiences and deep ecological knowledge. As custodians of a fragile and diverse ecosystem, the Marmas recognize the importance of conserving natural resources and mitigating environmental risks (Rahman et al., 2012). Their adaptive strategies, such as agroforestry and soil conservation, demonstrate a harmonious balance between human needs and ecological sustainability (Ahmed et al., 2015). These practices align with the principles of resilience theory, which emphasizes the capacity to adapt and transform in response to environmental changes (Walker et al., 2004).

Disasters, however, challenge the Marmas' resilience by exposing systemic vulnerabilities and socio-economic inequalities. The impacts of disasters are often compounded by marginalization and limited access to resources and services (Islam et al., 2014). For Indigenous communities like the Marmas, addressing these challenges requires a holistic approach that integrates social, cultural, and environmental dimensions of resilience (Adger, 2000). Efforts to enhance disaster preparedness and recovery must consider the unique needs and perspectives of indigenous populations to ensure equitable and sustainable outcomes (Smith et al., 2019).

The Marma indigenous communities' perspectives on resilience, spirituality, and environmental awareness offer valuable insights into the complex interplay between culture, environment, and disasters. Their experiences underscore the importance of integrating indigenous knowledge and values into disaster risk reduction and environmental sustainability initiatives. By recognizing and supporting the resilience of indigenous communities, policymakers and practitioners can foster more inclusive and effective responses to the challenges of climate change and environmental degradation. This study aims to contribute to this endeavor by shedding light on the rich cultural and spiritual heritage of the Marma people and their role in building a resilient and sustainable future (Table 6.1).

Findings

Spirituality and Connection to Nature Human existence is deeply intertwined with nature, yet modern lifestyles often alienate us from this intrinsic connection. Spirituality offers a bridge to restore harmony between humanity and the natural world, emphasizing reverence, coexistence, and sustainability. Across cultures and traditions, the spiritual relationship with nature has been a source of wisdom, guiding ethical behaviors and practices that nurture both the environment and the soul. This theme explores the profound link between spirituality and the natural world, delving into land-based perspectives, spiritual practices for sustainability, and the moral and ethical frameworks that advocate for environmental stewardship.

Table 6.1 Shows themes and subthemes resilience, spirituality, and environmental awareness in the face of disasters: perspectives from marma indigenous communities

Theme	Subtheme
Spirituality and connection to nature	– Land-based perspective on adaptation
	– Spiritual practices for sustainability
	– Morality and environmental ethics
Community and displacement	– Loss of indigenous land and identity
	– Community resilience in disaster
	– Breaking of social bonds
Environmental challenges and disasters	– Human-induced environmental damage
	– Impact of disasters on daily life
	– Mitigation and preparedness
Government and institutional roles	– Corruption and relief distribution
	– Need for education and awareness
	– Support for vulnerable communities
Cultural preservation amid challenges	– Threats to indigenous practices
	– Role of festivals and community gatherings
	– Land-based unity and respect

Land-Based Perspective on Adaption Nature, as the ultimate provider and sustainer, has always been revered for its inherent ability to adapt and support life. The concept of "land-based" refers to the naturally created elements of the earth, untouched by human fabrication. This perspective highlights the interconnectedness of all living beings and the essential role of nature in sustaining life. Land-based elements, such as trees and plants, offer invaluable resources. As one perspective articulates, "Land-based is naturally created creation, and which are not created by human beings. The contents of land-based help every living being. Some trees give medicine, some are food for other animals." This acknowledgment of nature's intrinsic value calls for a deeper appreciation and mindful interaction with the natural world.

Spiritual Practices for Sustainability Spiritual traditions have long recognized the importance of living in harmony with nature. One illustrative example is the story of Buddha achieving enlightenment under the Bodhi tree. This narrative underscores the mutual relationship between humans and nature—nature provides shelter and sustenance, and in return, humans are tasked with its preservation. As noted, "Buddha became a Buddha under a Bodhi tree...he wanted to give an example that the tree and nature have given you shelter, you have to save them." However, modern practices often contradict this ethos, as the act of cutting trees without replenishment has become commonplace. The wisdom of spiritual teachings emphasizes proactive care: "We cut five trees but never plant one. Buddha told to save the Bodhi tree, not to pray to it." Such teachings inspire actionable steps toward sustainable living, urging humanity to honor nature through tangible efforts.

Morality and Environmental Ethics Spirituality often provides a moral compass that extends beyond human interactions to encompass the treatment of the environment. Principles such as the Panchashila—non-killing, truthfulness, non-stealing, non-aggression, and abstinence from intoxication—reflect a code of conduct that fosters peace and respect for all forms of life.

"In my spirituality, Panchashila has been defined...do not kill, do not lie, do not steal, non-aggression, and not being intoxicated." These tenets are not only relevant to interpersonal relationships but also to the environment. For example, non-aggression extends to refraining from harming ecosystems, and abstinence from intoxication can include avoiding actions that pollute or degrade the environment. Furthermore, the centrality of peace in spiritual teachings resonates universally: "Peace is the main theme in every religion. But if there is no peace in the religion, what will we do with the religion?" This sentiment highlights the importance of aligning spiritual values with environmental ethics to achieve holistic peace.

The connection between spirituality and nature offers profound insights into how humanity can foster a sustainable and harmonious existence. By embracing land-based perspectives, practicing sustainability through spiritual teachings, and adhering to ethical principles, individuals and societies can cultivate a deeper respect for the natural world. In doing so, spirituality not only guides personal growth but also inspires collective action to protect and preserve the environment. This harmonious integration of spiritual values and environmental stewardship holds the key to a balanced and flourishing planet for generations to come.

Community and Displacement Communities form the backbone of human society, offering identity, support, and resilience in the face of challenges. However, displacement—whether due to economic, environmental, or social pressures—can disrupt these essential structures, leading to the loss of cultural heritage and social cohesion. This theme delves into the multifaceted impact of displacement on communities, focusing on the loss of indigenous land and identity, the resilience of communities in disasters, and the breaking of social bonds that underpin communal life.

Loss of Indigenous Land and Identity Displacement often results in the loss of ancestral lands, which are deeply tied to indigenous identity and culture. The forced or voluntary migration of Indigenous peoples disrupts their connection to the land and jeopardizes the preservation of their cultural heritage. One perspective captures this reality: "Bengalis settled here. They sold their land and went to remote areas like Ruma and Thanchi. It was their choice." While the act may have been voluntary, the consequences—loss of ancestral land—are profound. This displacement is further compounded by cultural erosion: "If our youngsters do not know about the language, this will be lost." The loss of language signifies a deeper loss of identity, as it disconnects future generations from their heritage and traditions. Preserving indigenous identity requires intentional efforts to maintain linguistic and cultural practices.

Community Resilience in Disaster Despite the challenges posed by displacement, communities often exhibit remarkable resilience, especially in the face of natural disasters. Collaborative efforts and external support can strengthen this resilience and foster a spirit of solidarity. As an example, "We built two cyclone centers with the help of KARITAS...later, the authority helped us to build four centers in this area." These centers represent a collective commitment to safeguarding lives and ensuring communal well-being. Additionally, the willingness to aid others during crises is evident: "In various disasters, they have helped us...we permit everyone in our centers." Such acts of generosity highlight the enduring strength of community bonds, even in the face of displacement and adversity.

Breaking of Social Bonds While resilience is a testament to community strength, displacement also has a detrimental impact on social cohesion. The breaking of social bonds often leaves communities fragmented and vulnerable. This fragmentation is poignantly expressed: "Community bondage has been broken here...life is tough, people are poor." Economic hardships and the struggle for survival exacerbate the breakdown of communal ties. Environmental changes further strain these bonds: "There was forest 50–60 years ago. Forest has been cut down. Population increased a lot." The loss of natural resources and the pressures of overpopulation erode traditional ways of life, making it difficult for communities to maintain their unity and identity.

The interplay between community and displacement reveals both the fragility and resilience of human connections. While displacement often results in the loss of indigenous land, identity, and social bonds, it also demonstrates the capacity of communities to adapt and rebuild in the face of adversity. Preserving cultural heritage, fostering resilience, and addressing the root causes of displacement are essential to sustaining the social fabric that binds communities together. By prioritizing these efforts, societies can mitigate the impact of displacement and nurture stronger, more cohesive communities for the future.

Environmental Challenges and Disasters Environmental challenges and disasters have profound implications for human life and the planet. While some of these challenges are natural, others are a result of human activity that exacerbates the vulnerability of ecosystems and communities. From the degradation of natural resources to the devastating impact of disasters on daily life, these issues demand immediate attention and action. This theme examines the role of human-induced environmental damage, the impact of disasters on daily life, and strategies for mitigation and preparedness to address these pressing concerns.

Subtheme 1: Human-Induced Environmental Damage

Human activities have significantly contributed to environmental degradation, impacting ecosystems and climate stability. Industrial processes, deforestation, and unregulated development are among the key contributors to this damage.

As one perspective notes, "Thermal power plantations have a bad impact on the climate. Wastages from different industries are coming out to air and water." The pollution of air and water resources poses severe health risks and disrupts ecological balance. In regions like Bandarban, the situation is further aggravated by unsustainable practices: "In Bandarban, people are cutting the trees, cutting the hills, and making roads." These activities not only threaten biodiversity but also increase the risk of landslides and soil erosion, endangering both the environment and human settlements.

Subtheme 2: Impact of Disasters on Daily Life

Natural disasters, often intensified by environmental degradation, have devastating consequences on daily life, displacing families and disrupting essential services. The immediate aftermath of such events highlights the vulnerability of affected communities.

One harrowing account illustrates the personal toll: "The roof of our house was blown away...we starved for three days after SIDR." Beyond the immediate physical destruction, disasters also have long-term effects on education and community well-being: "Students have been affected massively. They've lost their home, educational instruments, books—everything." These disruptions underscore the importance of building resilience and ensuring that basic needs, such as shelter, food, and education, are prioritized in disaster recovery efforts.

Subtheme 3: Mitigation and Preparedness

To address environmental challenges and reduce the impact of disasters, mitigation and preparedness are critical. Proactive measures, including community education and environmental conservation, play a vital role in minimizing risks.

Following disasters like SIDR, communities have learned the value of preparedness: "After SIDR, we taught people to store dry food and drinking water." Such practices can save lives and reduce suffering during emergencies. Additionally, environmental conservation efforts, such as planting trees, contribute to long-term resilience: "In Barisal, plant trees at the bank of the river so that soil erosion doesn't happen." These actions not only protect the environment but also safeguard livelihoods and infrastructure from future disasters.

Environmental challenges and disasters present significant threats to human life and the natural world. The intersection of human-induced damage and natural disasters calls for urgent, coordinated efforts to mitigate their impact. By addressing the root causes of environmental degradation, supporting disaster preparedness, and fostering community resilience, societies can better navigate these challenges. A commitment to sustainable practices and proactive planning is essential to protect the planet and ensure a safer, more equitable future for all.

Government and Institutional Roles Governments and institutions play a critical role in shaping the lives of vulnerable communities, particularly during times of crisis. Their responsibilities include ensuring effective resource distribution, promoting education and awareness, and supporting marginalized groups. However, these efforts are often undermined by corruption, inadequate planning, and insufficient outreach. This theme explores the challenges and opportunities associated with government and institutional roles, focusing on corruption in relief distribution, the need for education and awareness, and support for vulnerable communities.

Corruption and Relief Distribution Corruption significantly hampers the effectiveness of relief and support programs, leaving vulnerable communities underserved and disillusioned. Despite substantial government allocations, the benefits often fail to reach those in need. As one community member observed, "Government allocated 25 thousand crore taka for our area, but people got only 5000 taka." Such discrepancies highlight systemic inefficiencies and a lack of accountability. Additionally, the quality of distributed resources is frequently compromised: "We were supposed to get good quality plastic, but they gave us poor quality plastic." Addressing corruption and ensuring transparency in relief efforts are essential to build trust and effectively support affected populations.

Need for Education and Awareness Education and awareness are vital for empowering communities and preserving cultural identities. Yet, these areas are often neglected, especially in marginalized regions. "No one thought about our language... we need teachers for our children," one individual emphasized, pointing to the lack of educational infrastructure for indigenous groups. Awareness campaigns tailored to specific communities can also play a crucial role in addressing local challenges: "You have to send area-based awareness-building messages." Strengthening education and communication systems can help bridge knowledge gaps and foster community resilience.

Support for Vulnerable Communities Governments and institutions have a moral obligation to prioritize support for vulnerable populations, including students and low-income families. While some initiatives exist, they are often insufficient or poorly implemented. "Our MP is very kind. Government must help students with shelter and essential commodities," reflects the need for targeted assistance. However, even basic provisions, such as food aid, are frequently mismanaged: "If the government gives us 30kg rice, we get only 25kg." Ensuring fair distribution and expanding support mechanisms are critical steps toward equitable development.

The role of government and institutions in supporting communities is indispensable, yet fraught with challenges. By addressing corruption, investing in education and awareness, and enhancing support for vulnerable groups, these entities can fulfill their mandate to uplift and empower marginalized populations. Transparent and accountable governance is essential to foster trust and deliver meaningful change.

Cultural Preservation Amid Challenges Cultural preservation is a cornerstone of identity and community cohesion, particularly for indigenous populations facing the pressures of modernization and displacement. Challenges such as the erosion of traditional practices, inadequate support for cultural education, and the need for unity in diversity threaten the survival of these rich heritages. This theme examines the threats to indigenous practices, the role of festivals and community gatherings, and the importance of land-based unity and respect in maintaining cultural integrity.

Threats to Indigenous Practices Modernization and external influences often lead to the decline of indigenous practices, resulting in a loss of cultural identity. This is exacerbated by the lack of institutional support for preserving native languages and traditions. "Less knowledge of their own land-based makes people convert to other religions," one perspective notes, emphasizing the role of cultural alienation in religious shifts. Furthermore, "Nobody ever thought about teaching our children in Marma language" underscores the urgent need for linguistic preservation to maintain cultural continuity. Revitalizing indigenous practices requires deliberate efforts to integrate them into educational systems and community programs.

Subtheme 2: Role of Festivals and Community Gatherings Festivals and community events serve as vital platforms for cultural expression and solidarity, yet they too face challenges such as resource mismanagement and shifting priorities. "During Prabarana, we are supposed to get 3 tons of rice, but one ton goes missing," highlights how corruption undermines cultural celebrations. Nevertheless, these gatherings also provide opportunities for education and resilience-building: "We try to make people understand how to stay safe during cyclones." Strengthening the role of festivals as cultural anchors can help reinforce community bonds and promote collective well-being.

Subtheme 3: Land-Based Unity and Respect Unity and mutual respect are fundamental to the survival of diverse cultural traditions. Land-based practices emphasize coexistence and learning from one another. "We try to live like brothers...in various disasters, they have helped us," reflects the communal spirit that sustains resilience during hardships. Moreover, "We have to learn about other religions. Peace is the main theme in land-based learning," highlights the importance of intercultural understanding in fostering harmony. Promoting land-based unity and mutual respect can strengthen the cultural fabric and create a foundation for lasting peace.

Cultural preservation amid modern challenges requires a multifaceted approach that addresses the erosion of indigenous practices, enhances the role of community events, and fosters unity and respect. By valuing and supporting cultural diversity, societies can ensure that these traditions continue to thrive and enrich the collective human experience. Intentional efforts to preserve cultural heritage will not only honor the past but also empower communities to navigate the future with confidence and resilience.

Discussion

The findings of this study underscore the profound interconnections between resilience, spirituality, and environmental awareness within Marma indigenous communities. These dimensions not only inform the Marmas' adaptive strategies in the face of disasters but also provide a framework for understanding broader Indigenous responses to socio-environmental challenges. This discussion elaborates on the key findings, emphasizing their implications for sustainable development, disaster risk reduction, and cultural preservation. Resilience in the Face of Disasters. The Marma communities exemplify resilience as a dynamic interplay between cultural identity, social networks, and traditional ecological knowledge. Their collective responses to environmental challenges highlight the importance of social capital in sustaining communal well-being during crises (Adger, 2000; Adger et al., 2005). For instance, the establishment of cyclone centers with community participation reflects their proactive approach to disaster preparedness (Islam & Walkerden, 2014). These centers, built with the support of organizations like KARITAS, illustrate how external assistance can amplify local resilience when aligned with community needs and values (Huq et al., 2013). However, resilience among the Marmas is not without its vulnerabilities. Disasters such as cyclones and landslides exacerbate existing socio-economic disparities and expose gaps in institutional support (Ahmed, Haque, & Neelormi, 2019). The Marmas' reliance on communal solidarity to overcome these challenges underscores the need for policies that strengthen local capacities while addressing systemic inequities (Rahman, 2020). Integrating Indigenous perspectives into national disaster management strategies can ensure that interventions are culturally appropriate and inclusive (Nakashima & Roué, 2002).

Spirituality as a Lens for Environmental Stewardship Spirituality is central to the Marmas' environmental ethics, guiding their interactions with nature and shaping their responses to ecological challenges. Their reverence for sacred groves and trees, for example, reflects a deep-seated commitment to biodiversity conservation (Hossain & Chakma, 2018). This spiritual ethos resonates with global environmental movements advocating for the intrinsic value of nature (Chakma, 2016). The findings also reveal the practical implications of Marma spirituality for sustainability. Rituals associated with planting and harvesting embody principles of reciprocity and stewardship, promoting practices that balance human needs with ecological integrity (Chakma & Chakma, 2021). Such traditions offer valuable insights for contemporary environmental policies, particularly in promoting sustainable land-use practices (Ahmed et al., 2015). Moreover, the Marmas' emphasis on peace and harmony, as encapsulated in the Panchashila principles, extends to their environmental interactions (Sopher, 1968). These principles advocate for non-aggression and respect for all life forms, fostering a holistic approach to environmental ethics. By aligning spiritual values with modern conservation efforts, policymakers can draw on indigenous wisdom to address the ecological crises of our time (Nakashima et al., 2000).

Environmental Awareness and Adaptive Strategies The Marmas' environmental awareness is rooted in their lived experiences and traditional knowledge. Their adaptive strategies, such as agroforestry and soil conservation, demonstrate a nuanced understanding of ecological processes (Berkes & Folke, 1998). These practices not only enhance food security but also contribute to climate resilience by mitigating soil erosion and preserving biodiversity (Rahman et al., 2012). However, the findings highlight the threats posed by human-induced environmental damage, including deforestation and unregulated development. In regions like Bandarban, the cutting of trees and hills for infrastructure projects jeopardizes both ecological stability and community resilience (Islam et al., 2014). Addressing these challenges requires a collaborative approach that integrates local knowledge with scientific expertise (Dumont et al., 2017). The Marmas' proactive measures, such as storing dry food and drinking water in anticipation of disasters, exemplify the importance of preparedness. These practices can inform broader disaster risk reduction strategies, emphasizing the role of community-based initiatives in enhancing resilience (Ahmed, Haque, & Neelormi, 2019). Supporting such efforts through capacity-building programs and resource allocation can strengthen the Marmas' ability to adapt to future challenges (Walker et al., 2004).

The Interplay of Displacement and Cultural Identity Displacement, whether driven by economic pressures or environmental crises, poses significant challenges to the Marmas' cultural identity and social cohesion. The loss of ancestral lands disrupts their connection to traditional practices and threatens the preservation of their language and heritage (Berkes et al., 2000). The findings emphasize the need for intentional efforts to safeguard indigenous cultures, including the inclusion of Marma language education in school curricula (Hossain & Chakma, 2018). Despite these challenges, the Marmas demonstrate remarkable resilience in maintaining their cultural identity. Festivals and community gatherings serve as vital platforms for cultural expression and solidarity, even in the face of resource constraints (Rahman, 2020). Strengthening the role of such events can reinforce communal bonds and foster a sense of belonging (Adger, 2000). The Marmas' experiences also highlight the impact of displacement on social cohesion. The breaking of social bonds, compounded by economic hardships and environmental changes, underscores the importance of addressing the root causes of displacement (Adger et al., 2005). Policies that promote equitable access to resources and support sustainable livelihoods can mitigate the adverse effects of displacement on community well-being (Ahmed et al., 2015).

Implications for Policy and Practice The findings of this study have several implications for policymakers, practitioners, and researchers working in the fields of disaster management, environmental conservation, and cultural preservation. First, integrating Indigenous perspectives into disaster risk reduction strategies can enhance their effectiveness and cultural relevance (Nakashima et al., 2000). For instance, recognizing the Marmas' traditional knowledge and spiritual values can inform the design of interventions that are both sustainable and respectful of local contexts (Hossain & Chakma, 2018). Second, promoting education and awareness within Indigenous communities is critical for empowering them to navigate contemporary challenges. Initiatives that prioritize linguistic and cultural preservation can strengthen community resilience by fostering a sense of identity and belonging (Smith et al., 2019). Additionally, area-specific awareness campaigns can address local environmental and disaster-related issues, building community capacities to respond to crises (Islam & Walkerden, 2014). Third, addressing systemic issues such as corruption in relief distribution and resource mismanagement is essential for ensuring that government and institutional support reaches vulnerable populations. Transparent governance and accountability mechanisms can build trust and enhance the effectiveness of relief efforts (Rahman et al., 2012). Finally, fostering partnerships between Indigenous communities, civil society organizations, and government agencies can amplify the impact of resilience-building initiatives (Adger, 2000). Collaborative approaches that leverage the strengths of each stakeholder can create synergies for sustainable development and disaster preparedness (Huq et al., 2013).

The Marma indigenous communities' perspectives on resilience, spirituality, and environmental awareness offer valuable insights into the complex interplay between culture, environment, and disasters. Their experiences underscore the importance of integrating traditional knowledge and spiritual values into contemporary approaches to sustainability and disaster management (Chakma & Chakma, 2021). By recognizing and supporting the resilience of indigenous communities, policymakers and practitioners can foster more inclusive and effective responses to the challenges of climate change and environmental degradation (Ahmed, Mallick, & Rahman, 2019). This discussion highlights the need for a holistic approach that addresses the socio-cultural, environmental, and institutional dimensions of resilience (Berkes & Folke, 1998). By prioritizing equity, inclusivity, and cultural preservation, societies can create a foundation for sustainable and harmonious coexistence. The lessons learned from the Marma communities can serve as a guide for building resilient futures in the face of growing environmental and socio-economic challenges (Walker et al., 2004).

Chapter 7
Land-Based Spirituality and Climate Adaptation in Khyeng Indigenous Communities

This chapter explores the interplay between land-based spirituality and climate adaptation strategies within the Khyeng Indigenous communities of the Chittagong Hill Tracts in Bangladesh. Anchored in deep cultural and ecological traditions, the Khyeng demonstrate resilience amidst the profound impacts of human-induced climate change. The findings highlight critical challenges, including disruptions to agricultural livelihoods, health crises during disasters, and economic vulnerabilities. Despite these adversities, the Khyeng sustain adaptive strategies rooted in their spiritual practices, such as rituals for environmental harmony, daily gratitude, and seasonal agricultural preparations. These cultural mechanisms are complemented by community-driven initiatives like disaster preparedness and advocacy for land rights, underscoring the necessity of integrating Indigenous knowledge into broader climate policies. This chapter underscores the interconnectedness of environmental, socio-economic, and cultural dimensions in fostering sustainable adaptation. By valuing the Khyeng\u2019s spiritual and ecological knowledge, policymakers can advance inclusive and resilient approaches to climate action, providing lessons applicable to global efforts in environmental justice and sustainability.

The global climate crisis, largely driven by human activities such as deforestation, industrialization, and greenhouse gas emissions, has had profound effects on vulnerable communities worldwide. Among these, the Khyeng Indigenous communities in Bangladesh, situated in the Chittagong Hill Tracts, stand out for their unique land-based spirituality and deep ecological knowledge. These communities offer a powerful lens through which to explore sustainable climate adaptation strategies rooted in Indigenous worldviews. This document examines the intersection of human-induced climate change and land-based spirituality, emphasizing how the Khyeng people navigate and adapt to environmental changes through their cultural and spiritual practices.

R. Datta, *Decolonizing Climate Adaptation*,
https://doi.org/10.1007/978-981-96-9140-1_7

Human-Created Climate Crisis

The Anthropocene epoch is characterized by significant human impacts on the Earth's ecosystems (Crutzen, 2006). Activities such as deforestation, unsustainable agriculture, and industrial development have contributed to global warming, biodiversity loss, and ecosystem degradation (IPCC, 2021). In Bangladesh, these effects are particularly acute, given the country's geographic location and socio-economic vulnerabilities (Huq et al., 2015).

The Chittagong Hill Tracts (CHT) have witnessed deforestation, soil erosion, and changing rainfall patterns due to human activities (Chakma, 2018). These impacts exacerbate the challenges faced by Indigenous communities, disrupting traditional agricultural practices and threatening food security (Alam et al., 2017). Moreover, infrastructure projects, such as dam construction and logging, have displaced Indigenous populations and degraded their ancestral lands (Roy et al., 2020).

Despite these challenges, Indigenous communities have demonstrated remarkable resilience. Their adaptation strategies are deeply intertwined with their cultural and spiritual practices, which provide a framework for understanding and responding to environmental changes (Berkes, 2018).

Land-Based Spirituality in Khyeng Communities

Land-based spirituality is central to the Khyeng people's identity and way of life. Their spiritual beliefs are rooted in the interconnectedness of humans, land, and nature, reflecting a holistic worldview that prioritizes harmony with the environment (Chakma & Tripura, 2019). Sacred groves, ancestral lands, and natural elements such as rivers and forests are revered as embodiments of spiritual entities, guiding community practices and decision-making (Tanchangya, 2020).

Spiritual Practices and Climate Adaptation

1. Sacred forests and conservation the Khyeng maintain sacred forests, which serve as vital repositories of biodiversity and cultural heritage. These forests act as natural buffers against climate impacts, such as floods and landslides, while providing resources for traditional medicine and rituals (Roy et al., 2018).
2. Rituals and seasonal indicators rituals tied to agricultural cycles reflect the Khyeng's deep ecological knowledge. Seasonal indicators, such as the flowering of certain plants or the behavior of wildlife, inform planting and harvesting practices (Tanchangya, 2020). These practices enhance food security and resilience to climatic variability (Berkes, 2018).
3. Community gatherings and knowledge transmission spiritual ceremonies serve as platforms for intergenerational knowledge-sharing, ensuring the transmission

of ecological wisdom and cultural values. This communal approach fosters collective action and strengthens social cohesion, essential for climate adaptation (Chakma, 2018).

Challenges to Land-Based Spirituality

The Khyeng's land-based spirituality faces significant threats due to external pressures. Encroachment by non-Indigenous settlers, deforestation, and commercialization of natural resources disrupt sacred sites and undermine traditional practices (Alam et al., 2017). Additionally, climate-induced displacement poses existential risks to the Khyeng's cultural identity, as losing access to ancestral lands severs their spiritual connection to the environment (Chakma & Tripura, 2019).

Climate Adaptation Strategies

The Khyeng's climate adaptation strategies are informed by their land-based spirituality and ecological knowledge. These strategies offer valuable insights for sustainable and inclusive climate policies.

1. Agroforestry and Sustainable Agriculture

   Combining traditional knowledge with modern techniques, the Khyeng practice agroforestry, which integrates trees and crops to enhance soil fertility and biodiversity (Chakma, 2018). This approach mitigates the impacts of deforestation and supports sustainable livelihoods (Huq et al., 2015).
2. Water Management

   The Khyeng's water management practices, such as constructing small-scale irrigation systems and preserving natural springs, address water scarcity and support agricultural productivity (Roy et al., 2020). These methods align with their spiritual emphasis on protecting water sources as sacred entities (Tanchangya, 2020).
3. Community-Based Resource Management

   The Khyeng's communal land tenure system promotes equitable resource distribution and sustainable use. This system strengthens community resilience by fostering collective responsibility for environmental stewardship (Berkes, 2018).
4. Advocacy and Collaboration

   Indigenous-led advocacy efforts, supported by NGOs and researchers, aim to secure land rights and protect sacred sites from exploitation (Alam et al., 2017). Collaborative projects that integrate Indigenous knowledge with scientific research enhance the effectiveness of climate adaptation initiatives (Huq et al., 2015).

Global Relevance and Policy Implications

The Khyeng's experiences highlight the importance of integrating Indigenous knowledge and spirituality into global climate strategies. Recognizing the rights of Indigenous communities and incorporating their perspectives into policy frameworks can enhance climate resilience and biodiversity conservation (IPCC, 2021). For example, implementing community-led conservation initiatives and providing legal protection for sacred sites can support both cultural preservation and environmental sustainability (Chakma & Tripura, 2019).

The human-created climate crisis poses unprecedented challenges to Indigenous communities like the Khyeng in Bangladesh. However, their land-based spirituality and ecological knowledge offer valuable lessons for climate adaptation. By respecting and supporting Indigenous worldviews, policymakers and practitioners can develop more inclusive and effective approaches to addressing climate change. The Khyeng's resilience underscores the need for collaborative efforts that honor cultural diversity and prioritize environmental justice (Table 7.1).

Findings

Impacts of Climate Change on Livelihoods and Health The impacts of climate change extend beyond the physical environment, deeply influencing the socio-economic and health conditions of the Khyeng indigenous communities. Dependent on nature for their livelihoods, the Khyeng face immense challenges as changing climatic patterns disrupt agricultural practices, exacerbate health crises, and intensify economic hardships. This theme delves into the tangible and interwoven effects of climate change on livelihoods and health, focusing on agricultural losses, health challenges during disasters, and economic hardships.

Table 7.1 shows themes and subthemes land-based spirituality and climate adaptation in khyeng indigenous communities

Theme	Subtheme
Impacts of climate change on livelihoods and health	– Agricultural losses – Health challenges during disasters – Economic hardships
Spiritual practices and land-based resilience	– Rituals for environmental harmony – Gratitude through daily practices – Seasonal preparations rooted in spirituality
Challenges of land tenure and rights	– Loss of ancestral land – Landless livelihoods – Advocacy for land rights
Community-driven climate adaptation strategies	– Disaster preparedness – Education and training – Building collective shelter and resources

Agricultural Losses Frequent floods, erratic rainfall, and prolonged heatwaves have wreaked havoc on the Khyeng's traditional farming practices. Paddy fields, once a source of sustenance and economic stability, are increasingly vulnerable to these climatic changes. As one farmer lamented, "This year, paddy fields were wasted. They had to harvest again… searching for seeds everywhere and buying at a high price." This quote encapsulates the cyclical hardships faced by the Khyeng: crop failures lead to food insecurity and rising costs for agricultural inputs, further straining already limited resources. Traditional farming methods, which depend on predictable weather patterns, are proving inadequate in the face of such variability, leaving farmers struggling to adapt. The loss of crops also has cascading effects on community well-being. Reduced yields not only jeopardize food security but also undermine the ability to sell surplus produce, affecting the overall economic health of families. Addressing these challenges necessitates the introduction of climate-resilient farming practices and improved access to agricultural support systems.

Health Challenges During Disasters Climate change has significantly disrupted access to healthcare in Khyeng communities, particularly during floods and other disasters. Vulnerable populations, such as children and pregnant women, bear the brunt of these disruptions. "The kids are getting sick… pregnant women are unable to have checkups or get proper treatment," observed a community elder. Floodwaters often render healthcare facilities inaccessible, while contaminated water sources exacerbate the spread of diseases such as diarrhea, malaria, and skin infections. Limited healthcare infrastructure and resources further compound the problem, leaving many without essential medical care during critical times. The health impacts of climate change are not restricted to physical ailments; mental health issues also emerge as communities grapple with the stress and uncertainty of recurring disasters. Strengthening healthcare systems, improving disaster preparedness, and ensuring access to clean water and sanitation are crucial steps to mitigate these challenges.

Subtheme 2.3: Economic Hardships

Economic vulnerability is another stark consequence of climate change, particularly for marginalized groups within Khyeng communities. Daily laborers, many of whom are women, face heightened economic insecurity as floods disrupt local economies and reduce employment opportunities. "Many women are daily laborers… their families' economic conditions have collapsed," shared a community leader. Job losses during and after disasters lead to financial instability, making it difficult for families to recover. The rising costs of basic necessities, such as seeds and food, further exacerbate these hardships, leaving families trapped in a cycle of poverty.

The economic fallout also has broader social implications, including increased dependence on external aid and the erosion of community self-reliance. To address these challenges, initiatives that provide alternative livelihoods, financial assistance, and skill-building opportunities are essential. Building economic resilience within the community can help mitigate the long-term impacts of climate-induced economic hardships.

Spiritual Practices and Land-Based Resilience For the Khyeng communities, spirituality is deeply interwoven with their relationship to the land and natural resources. This connection shapes their resilience, offering a framework for navigating environmental challenges through practices of respect, gratitude, and seasonal preparation. This theme explores how spiritual practices foster land-based resilience, focusing on rituals for environmental harmony, daily gratitude practices, and seasonal preparations rooted in spirituality.

Rituals for Environmental Harmony The Khyeng's spiritual worldview emphasizes accountability to nature, recognizing the interconnectedness of all living beings. Rituals aimed at restoring environmental harmony reflect this ethos, fostering a culture of respect and stewardship. "We must ask for forgiveness… as we destroyed their house to build ours," said a community elder, highlighting a ritual of reconciliation with nature. These practices acknowledge the impact of human actions on ecosystems and serve as a moral compass, guiding sustainable interactions with the environment. Such rituals underscore the potential of spiritual traditions to inspire environmental responsibility and collective action in the face of ecological crises.

Gratitude Through Daily Practices Daily expressions of gratitude are central to the Khyeng's spiritual connection with the land. Evening prayers, for example, serve as moments of thanksgiving for the resources provided by nature and as a way to seek blessings for future well-being. "Evening prayers are thanksgiving… people pray that the upcoming day can be fruitful and good," shared a villager. These practices foster a sense of humility and interdependence, reinforcing the community's commitment to sustainable living. By maintaining a daily spiritual connection to their environment, the Khyeng sustain a resilient mindset that helps them adapt to and endure environmental challenges.

Seasonal Preparations Rooted in Spirituality Agricultural practices among the Khyeng are closely tied to spiritual observations and seasonal cycles. Decisions about planting and harvesting are guided by rituals and an intuitive understanding of environmental conditions. "In lower lands, planting seeds should only be done according to the situation," explained a community farmer. This integration of spirituality and agricultural knowledge reflects a holistic approach to land management, ensuring that practices align with the rhythms of nature. Such seasonal preparations not only optimize productivity but also reinforce the community's spiritual bond with the land. The integration of spirituality into practical decision-making highlights the potential of indigenous knowledge systems to inform sustainable practices. Recognizing and supporting these traditions can contribute to broader efforts in climate adaptation and resilience.

The Khyeng indigenous communities offer a profound example of resilience through their spiritual practices and observations of climate change. While climate change poses significant challenges to their livelihoods, health, and economic stability, their spiritual connection to the land provides a foundation for navigating these difficulties. Rituals for environmental harmony, daily gratitude practices, and seasonal preparations rooted in spirituality demonstrate how cultural traditions can foster resilience and sustainability.

Addressing the impacts of climate change on the Khyeng requires integrating their spiritual and ecological knowledge into broader strategies for adaptation and mitigation. Strengthening healthcare systems, supporting sustainable livelihoods, and preserving cultural practices are essential components of this effort. By valuing the Khyeng's land-based spirituality, policymakers and practitioners can develop more inclusive and effective approaches to building resilience in the face of environmental change.

Challenges of Land Tenure and Rights Indigenous communities around the world face significant challenges regarding land tenure and rights, which are compounded by the impacts of climate change. These issues not only disrupt traditional ways of life but also hinder efforts toward sustainable development and environmental restoration. This document delves into two interconnected themes: the challenges of land tenure and rights, and community-driven climate adaptation strategies. These themes explore the loss of ancestral lands, landless livelihoods, advocacy for land rights, disaster preparedness, education and training, and the development of collective shelters and resources. Together, they highlight the resilience and agency of Indigenous communities in addressing these pressing concerns.

Loss of Ancestral Land "My father had 17 Kani of land but sold it to the mainstream community at a very cheap rate due to poverty." This statement reflects a harsh reality faced by many Indigenous families. Economic hardship often forces families to sell their ancestral lands, leading to significant displacement. Ancestral lands hold immense cultural, spiritual, and ecological significance, serving as the foundation for traditional knowledge systems and practices. The loss of these lands disrupts community cohesion, erodes cultural heritage, and exacerbates vulnerabilities. The primary driver of these land losses is poverty, a pervasive issue within Indigenous communities. Limited access to education, healthcare, and sustainable employment opportunities traps families in cycles of economic instability. As a result, selling land becomes a desperate measure to address immediate needs, even though it undermines long-term well-being. This dispossession has intergenerational impacts, leaving younger generations without the resources to sustain traditional livelihoods or advocate effectively for their rights.

Addressing this challenge requires systemic solutions that tackle poverty and provide Indigenous families with the means to retain their lands. Policies that support economic development while respecting cultural practices can help mitigate land loss. For instance, community-led initiatives that promote sustainable agriculture, eco-tourism, or artisanal crafts can generate income while preserving cultural heritage.

Landless Livelihoods "Many live on others' land… Forestation isn't possible here." This quote underscores the precarious situation of landless Indigenous families. Without land ownership, they are unable to engage in sustainable practices such as reforestation or traditional farming. This lack of autonomy hinders efforts to restore degraded ecosystems and adapt to the changing climate. Landlessness also limits access to resources such as water and timber, which are essential for traditional livelihoods. As a result, families are forced to seek employment in low-wage sectors or migrate to urban areas, where they face additional challenges such as discrimination and lack of social support. This displacement disrupts community networks and erodes cultural identities, further marginalizing Indigenous populations. To address landlessness, governments and organizations must prioritize land redistribution and tenure security. Providing legal recognition of Indigenous land rights can empower communities to manage their resources sustainably. Collaborative initiatives that involve Indigenous leaders, government agencies, and non-profits can create pathways for equitable land access and stewardship.

Advocacy for Land Rights "Government can give back our land by a proper procedure through changing the Constitution." Advocacy for land rights is a cornerstone of Indigenous resilience and survival. Restoring ancestral lands is not only a matter of justice but also a critical step toward climate adaptation. Secure land tenure enables communities to implement traditional ecological knowledge and practices that enhance biodiversity and resilience. Successful advocacy efforts often involve grassroots mobilization, legal action, and international support. For example, Indigenous communities worldwide have used platforms such as the United Nations Declaration on the Rights of Indigenous Peoples (UNDRIP) to assert their rights and demand government accountability. These movements highlight the need for systemic changes, including constitutional amendments and the establishment of legal frameworks that recognize Indigenous land ownership. Education and capacity-building are essential components of effective advocacy. By equipping community members with knowledge about their rights and tools for negotiation, advocacy efforts can gain traction and achieve meaningful outcomes. Furthermore, partnerships with allies in academia, civil society, and international organizations can amplify Indigenous voices and drive policy changes.

Theme 5: Community-Driven Climate Adaptation Strategies

Disaster Preparedness "Basic medicines for fever, cough, diarrhea... should be stored for emergencies." Disaster preparedness is a critical aspect of community-driven climate adaptation. Indigenous communities, particularly those in vulnerable regions, are often the first to experience the impacts of climate change, such as floods, droughts, and cyclones. Preparing for these disasters requires a proactive approach that includes storing essential supplies, developing evacuation plans, and building local capacity. Community-led initiatives play a vital role in enhancing disaster preparedness. For example, training programs that teach first aid, emergency response, and resource management can empower community members to respond effectively during crises. Additionally, incorporating traditional knowledge into disaster planning can enhance resilience. Indigenous practices such as water harvesting, seed preservation, and forest management provide valuable insights into sustainable adaptation. Governments and organizations can support these efforts by providing resources, technical assistance, and funding for community-driven projects. Strengthening infrastructure, such as building resilient shelters and improving communication networks, is also essential to ensure timely responses during emergencies.

Education and Training "Government should give people training... about types of disasters and what to do." Education and training are pivotal for building community resilience to climate change. Knowledge-sharing initiatives that focus on disaster risk reduction, sustainable practices, and environmental conservation can empower communities to adapt effectively. Formal and informal education systems can play complementary roles in this process. Schools can integrate climate education into their curricula, while community workshops can provide practical training tailored to local needs. For instance, teaching farmers about crop diversification, water management, and agroforestry can enhance food security and reduce vulnerability to climate shocks.

Collaborations between Indigenous communities, governments, and non-governmental organizations can create comprehensive training programs that address both immediate and long-term adaptation needs. These partnerships should prioritize culturally relevant approaches that respect Indigenous knowledge systems and practices.

Building Collective Shelter and Resources "Primary schools are used as shelters... but more needs to be done for displaced families." Developing robust community shelters and resources is crucial for protecting vulnerable populations during disasters. While schools and other public buildings often serve as temporary shelters, they are typically insufficient to accommodate the needs of displaced families.

Investing in dedicated disaster shelters that are resilient to extreme weather events can significantly enhance community safety. These shelters should be equipped with essential supplies, such as food, water, and medical kits, and designed to accommodate diverse needs, including those of children, elderly individuals, and persons with disabilities.

Community resource centers can also serve as hubs for adaptation initiatives, providing spaces for training, knowledge-sharing, and coordination. By fostering collaboration and mutual support, these centers can strengthen community cohesion and resilience.

The challenges of land tenure and rights, coupled with the need for community-driven climate adaptation strategies, highlight the interconnected nature of social, economic, and environmental issues faced by Indigenous communities. Addressing these challenges requires a holistic approach that prioritizes land rights, poverty alleviation, education, and infrastructure development. Indigenous resilience and agency are central to overcoming these obstacles. By recognizing and supporting their knowledge systems, governments and organizations can foster sustainable and inclusive solutions. Ultimately, empowering Indigenous communities to lead the way in addressing these issues not only benefits them but also contributes to global efforts to build a more equitable and resilient future.

Discussion

The findings reveal intricate relationships between climate change impacts, land-based spirituality, and adaptive strategies within Khyeng Indigenous communities. This discussion integrates the findings with existing literature to contextualize their significance and explore broader implications for climate adaptation and cultural resilience. By examining the Khyeng's experiences, we gain a deeper understanding of how Indigenous knowledge systems can inform sustainable responses to climate change.

Impacts of Climate Change on Livelihoods and Health

Agricultural Losses

The findings highlight significant agricultural disruptions due to climate change, consistent with broader studies on the vulnerability of Indigenous agricultural systems (Altieri, 2004). Erratic rainfall, frequent floods, and prolonged heatwaves in the Chittagong Hill Tracts (CHT) have destabilized traditional farming practices, similar to challenges faced by Indigenous farmers in other regions (Adger et al., 2005). Losses in paddy yields not only jeopardize food security but also undermine economic stability. This aligns with Khan et al. (2021), who emphasize the cascading effects of agricultural failures on household incomes and community well-being.

The cyclical nature of these hardships, as noted in the findings, underscores the need for climate-resilient farming techniques. Agroecological approaches, such as those suggested by Berkes (2018), could be integrated with Indigenous knowledge to improve resilience. Practices like crop diversification and water-efficient irrigation systems offer practical solutions while respecting cultural traditions (Pretty & Smith, 2004).

Health Challenges During Disasters

Health challenges, particularly during disasters, resonate with global patterns observed in climate-vulnerable communities. The findings highlight disruptions to healthcare access and the prevalence of waterborne diseases, echoing the observations of Watts et al. (2018) on climate-induced health crises. Vulnerable populations, such as children and pregnant women, face heightened risks, reflecting broader gendered and age-specific vulnerabilities documented by Ford and Pearce (2010).

Strengthening healthcare infrastructure in the CHT could mitigate these impacts. Mobile health clinics and telemedicine initiatives, as explored by Huq et al. (2015), offer viable solutions for remote communities. Additionally, integrating Indigenous healing practices with modern healthcare systems, as advocated by Jackson and Moggridge (2019), could enhance cultural sensitivity and effectiveness.

Economic Hardships

Economic hardships resulting from climate disruptions are a recurring theme in climate literature (Kumar et al., 2021). For the Khyeng, floods and agricultural losses reduce employment opportunities, disproportionately affecting women engaged in daily labor. This mirrors findings by Agrawal (2010), who discusses the intersection of climate vulnerability and socio-economic marginalization.

Developing alternative livelihoods is crucial to breaking the cycle of poverty. Initiatives like microfinance programs and skill-building workshops, as described by Alam et al. (2017), could empower women and other marginalized groups. Such efforts must prioritize equity and sustainability to address structural inequalities effectively (Schipper et al., 2020).

Spiritual Practices and Land-Based Resilience

Rituals for Environmental Harmony

The Khyeng's rituals for environmental harmony reflect a deep ecological consciousness, aligning with the concept of "spiritual ecology" discussed by Sponsel (2024). These practices foster accountability to nature and inspire sustainable behaviors. For example, rituals of reconciliation with nature resonate with

Indigenous philosophies worldwide that emphasize interconnectedness and reciprocity (Nelson & Wilson, 2018).

Supporting such practices could enhance community-driven conservation efforts. Collaborative projects that integrate spiritual traditions with modern environmental science, as proposed by Pretty and Smith (2004), offer a holistic approach to ecological restoration. Recognizing the cultural dimensions of sustainability could also strengthen policy frameworks, as suggested by Posey (1999).

Gratitude Through Daily Practices

Daily expressions of gratitude, such as evening prayers, reinforce the Khyeng's commitment to sustainable living. This aligns with studies highlighting the psychological and social benefits of gratitude practices in fostering resilience (Emmons & McCullough, 2003). The Khyeng's spiritual connection to the land provides a moral compass that guides their adaptation strategies, similar to findings by Berkes (2018) on Indigenous knowledge systems.

These practices could inform educational programs aimed at promoting sustainability. Incorporating spiritual values into environmental education, as explored by Wals et al. (2014), could cultivate a deeper sense of stewardship among younger generations.

Seasonal Preparations Rooted in Spirituality

The integration of spirituality and agricultural decision-making among the Khyeng reflects a holistic approach to land management. This resonates with Altieri (2004), who emphasizes the importance of traditional ecological knowledge in adapting to climatic variability. The Khyeng's seasonal preparations optimize resource use and align with natural cycles, showcasing the value of Indigenous knowledge in sustainable agriculture.

Promoting such practices through community-based adaptation programs could enhance resilience. Collaborative research initiatives, as suggested by Huq et al. (2015), could document and scale up successful Indigenous practices, ensuring their integration into broader climate strategies.

Challenges of Land Tenure and Rights

Loss of Ancestral Land

The loss of ancestral lands, driven by economic hardships, highlights the structural vulnerabilities faced by the Khyeng. This aligns with observations by Roy et al. (2018) on the erosion of Indigenous land rights in Bangladesh. Land dispossession disrupts cultural continuity and exacerbates economic marginalization, as documented by Chakma and Tripura (2019).

Policy interventions must address the root causes of land loss, such as poverty and legal ambiguities. Initiatives like community land trusts and legal aid programs, as explored by Alden Wily (2011), could empower Indigenous communities to reclaim and protect their lands. International frameworks like the United Nations Declaration on the Rights of Indigenous Peoples (UNDRIP) offer valuable tools for advocacy (Anaya, 2009).

Landless Livelihoods

Landlessness among the Khyeng limits their ability to engage in sustainable practices, reflecting broader challenges faced by landless Indigenous populations worldwide (Deininger & Feder, 2001). Without secure land tenure, efforts to restore ecosystems or practice traditional farming are severely constrained.

Collaborative land management models, such as participatory forestry programs, could address these challenges. Examples from Nepal's community forestry initiatives, as discussed by Acharya (2002), demonstrate the potential for equitable resource management and ecological restoration.

Advocacy for Land Rights

Advocacy for land rights is essential for the Khyeng's resilience and survival. The findings underscore the importance of grassroots mobilization and constitutional reforms, echoing global movements for Indigenous land rights (Anaya, 2009). Empowering communities through education and capacity-building, as highlighted by Schilling-Vacaflor (2017), can amplify their voices and drive policy changes.

Community-Driven Climate Adaptation Strategies

Disaster Preparedness

Community-led disaster preparedness initiatives align with findings by Mercer et al. (2010) on the effectiveness of participatory approaches. The Khyeng's emphasis on self-reliance, such as storing basic medicines, reflects adaptive strategies observed in other Indigenous communities (Carter et al., 2014).

Supporting these efforts through technical assistance and funding could enhance their impact. Integrating traditional knowledge with modern disaster management frameworks, as suggested by Gaillard et al. (2007), could improve resilience and reduce vulnerabilities.

Education and Training

Education and training are pivotal for building resilience, consistent with findings by Leal Filho et al. (2018). The Khyeng's emphasis on knowledge-sharing and disaster education reflects global best practices in community-based adaptation. Tailoring training programs to local contexts, as advocated by Schipper et al. (2020), could maximize their effectiveness.

Building Collective Shelter and Resources

Developing robust shelters and resource centers is critical for protecting vulnerable populations. The findings resonate with studies by Cannon et al. (2003) on the importance of infrastructure in reducing disaster risks. Collaborative planning and community engagement, as explored by Twigg (2004), could ensure that shelters meet diverse needs and enhance social cohesion.

Conclusion

The findings underscore the resilience and agency of the Khyeng Indigenous communities in navigating climate challenges. By integrating land-based spirituality with adaptive strategies, they offer valuable lessons for sustainable and inclusive climate responses. Existing literature supports the potential of Indigenous knowledge systems to inform global efforts in climate adaptation and resilience. Strengthening healthcare, supporting sustainable livelihoods, securing land rights, and promoting community-driven initiatives are essential steps toward building a more equitable and resilient future.

Chapter 8
Land-Based Spirituality as Climate Resilience in Manipuri Indigenous Communities

This chapter explores the transformative role of land-based spirituality in addressing climate resilience within Manipuri Indigenous communities. The primary goal is to illustrate how spiritual, cultural, and ecological practices sustain community well-being while offering a critique of mainstream Western environmental management. The chapter begins with an introduction highlighting the limitations of Western approaches, followed by findings emphasizing Indigenous practices. Themes such as sacred connections to land, rituals, and observations of climate change are discussed in depth, demonstrating their practical and spiritual relevance. Strategies for preserving Indigenous knowledge systems and cultural adaptation conclude the chapter, offering actionable insights for integrating these practices into broader environmental policies. This work aims to inspire global recognition of Indigenous contributions to sustainability and climate justice.

Mainstream Western environmental management often views nature as a commodity, leading to unsustainable resource extraction practices. As Robbins (2012) notes, this anthropocentric perspective undermines ecosystems' regenerative capacities. For example, large-scale deforestation and mining projects disrupt habitats, driving species to extinction (Wilson, 2016). Manipuri Indigenous communities have observed these impacts firsthand, reporting the loss of native plant species critical to their cultural and ecological systems (Datta et al., 2024). The Western environmental management practices, rooted in industrial and capitalist paradigms, have often prioritized economic growth over ecological sustainability. This approach, characterized by resource extraction, monoculture farming, and technocratic solutions, has created numerous environmental challenges, including biodiversity loss, climate change, and social inequities. By contrast, Manipuri Indigenous communities emphasize land-based spirituality, which offers holistic climate solutions grounded in harmony with nature. This section explores the inadequacies of Western environmental management and illustrates how the spiritual practices of the Manipuri Indigenous communities serve as a model for sustainable environmental stewardship.

R. Datta, *Decolonizing Climate Adaptation*,
https://doi.org/10.1007/978-981-96-9140-1_8

The adoption of monoculture agriculture, a hallmark of Western environmental practices, has led to significant soil degradation and reduced biodiversity (Altieri, 2004). This approach prioritizes yield over ecological balance, stripping soils of nutrients and leaving them vulnerable to erosion (Tilman et al., 2002). Manipuri farmers have resisted monoculture, emphasizing diverse cropping systems and rituals that honor the land's fertility, preserving its health for future generations (Singh et al., 2020). Technocratic approaches to climate change, such as geoengineering and carbon trading, often ignore the socio-cultural dimensions of environmental challenges (Hamilton, 2013). These solutions focus on mitigating symptoms rather than addressing root causes, perpetuating ecological imbalance. Manipuri Indigenous communities critique these approaches for their disconnection from the lived realities of marginalized populations who bear the brunt of climate impacts (Bhadra et al., 2021).

Land-based spirituality in Manipuri Indigenous communities reflects a deep-seated belief in the sacredness of nature. This worldview fosters a reciprocal relationship between humans and the environment, emphasizing respect and stewardship (Kimmerer, 2013). For instance, rituals such as the *Umang Lai Haraoba*, which honor forest deities, underscore the interconnectedness of all living beings (Chongtham et al., 2018). These practices not only sustain ecological balance but also instill a sense of accountability, which contrasts sharply with exploitative Western approaches (Datta, 2022). Land-based spirituality informs agricultural practices that prioritize sustainability over short-term gains. Manipuri farmers engage in rituals to honor the land, such as offering fruits to field spirits, ensuring a harmonious relationship with the environment. These practices enhance soil fertility, reduce dependency on chemical inputs, and align with principles of regenerative agriculture (Sharma et al., 2017).

Manipuri Indigenous communities have developed community-led adaptation strategies rooted in traditional knowledge and spiritual values. For example, they have established sacred groves that act as biodiversity hotspots and climate refuges (Khumbongmayum et al., 2005). These initiatives demonstrate the potential of Indigenous-led conservation efforts to complement mainstream environmental policies (Gadgil et al., 1993). The spiritual practices of Manipuri Indigenous communities offer critical lessons for global environmental governance. Recognizing the limitations of Western approaches, policymakers and environmentalists must integrate Indigenous knowledge systems into climate strategies (Whyte, 2018).

Indigenous land-based spirituality emphasizes the interconnectedness of ecological, social, and spiritual dimensions. This holistic perspective can inform sustainability frameworks that prioritize ecological integrity alongside human well-being (Berkes, 2008).

Empowering Indigenous communities as stewards of their ancestral lands is essential for effective climate action. Studies show that Indigenous-managed territories often outperform protected areas in biodiversity conservation (Chapin et al., 2009). Supporting Manipuri initiatives such as sacred grove preservation can serve as a model for co-management strategies (Kothari et al., 2013).

Integrating Indigenous perspectives into environmental education fosters cross-cultural understanding and respect for diverse worldviews (Smith, 1999). The Manipuri emphasis on gratitude and reciprocity can inspire global environmental movements to adopt more ethical and inclusive practices (Cajete, 2000). The challenges posed by mainstream Western environmental management underscore the urgent need for alternative approaches. Manipuri Indigenous communities, through their land-based spirituality and sustainable practices, offer invaluable insights into building resilient ecosystems and fostering climate justice. By embracing these perspectives, the global community can move toward a more equitable and sustainable future. Recognizing and valuing Indigenous contributions is not only a moral imperative but also a practical necessity for addressing the complex environmental challenges of our time.

Findings

Indigenous communities worldwide have long demonstrated unique ways of adapting to and mitigating the impacts of environmental changes. Among the Manipuri Indigenous communities, land-based spirituality emerges as a profound framework for climate resilience. This chapter explores how spiritual practices rooted in sacred connections to the land contribute to sustainable living, environmental stewardship, and climate adaptation. The themes and subthemes below delve into the intricate ways these communities harmonize their spiritual beliefs with ecological practices (Table 8.1).

Sacred Connection to Land and Spiritual Practices The Manipuri Indigenous worldview revolves around an intrinsic connection to the land. This bond is not merely utilitarian but deeply spiritual, reflecting a relationship of respect, reverence, and reciprocity with nature. The following subthemes highlight the integral role of spiritual practices in fostering this sacred connection.

Rituals for Land-based Knowledge Rituals serve as a means to acknowledge and respect the non-human entities that share the ecosystem. These ceremonies embody the Manipuri belief in coexistence with nature—a philosophy crucial for ecological balance and climate resilience. One community member reflects: "We must ask for forgiveness from them as we have destroyed their house to build ours… This puja helped me to create respect for the other living things." Such rituals involve prayers and offerings to displaced beings, including spirits of trees and animals, symbolizing the community's recognition of the interconnectedness of all life forms. This practice fosters a sense of accountability and mindfulness, ensuring that development or resource use does not disrupt the harmony of the environment. By integrating these rituals into daily life, the Manipuri communities promote values of coexistence and environmental stewardship. These practices not only reaffirm their sacred bond with the land but also cultivate a collective consciousness to mitigate harm to the environment, embodying resilience against climate disruptions.

Table 8.1 Shows themes and subthemes land-based focusing on land-based spirituality as climate resilience in Manipuri indigenous communities

Theme	Subthemes	Key Points
Sacred connection to land and spiritual practices	Rituals for land-based knowledge	Acknowledges non-human entities, fostering respect and coexistence
	Gratitude practices	Daily prayers and rituals express appreciation for the environment
	Rituals for agricultural sustainability	Ensures harmony with land spirits and promotes eco-friendly farming methods
Observations of climate change and environmental shifts	Decreasing rainfall and water scarcity	Observations of reduced rainfall impacting agriculture and water cycles
	Loss of biodiversity	Decline in species vital for ecological and cultural resilience
	Heatwaves and changing seasons	Increased temperatures disrupting agriculture and community well-being
Decline of indigenous knowledge systems	Loss of traditional practices	Erosion of culinary and medicinal knowledge
	Intergenerational knowledge gaps	Challenges in passing down traditional wisdom
	Dependence on modern alternatives	Shift toward industrial methods, overshadowing sustainable practices
Architectural and cultural sustainability	Traditional housing techniques	Eco-friendly homes built with renewable materials like bamboo
	Cultural symbolism in infrastructure	Rituals and symbols align construction with spiritual values
	Rituals for building harmony	Practices honor displaced creatures, maintaining spiritual balance
Strategies for cultural and climate adaptation	Model indigenous villages	Combines tradition with modernity for cultural preservation and climate resilience
	Documentation of indigenous knowledge	Preserving practices through recording and archiving
	Integrating education with indigenous values	Inclusion of indigenous principles in educational curricula

Gratitude Practices Gratitude practices are another cornerstone of Manipuri spirituality. Daily prayers and rituals acknowledge the benevolence of the land and its resources. As one participant eloquently stated: "The evening prayer is basically a thanksgiving prayer... people pray that the upcoming day can be fruitful and good like today." These prayers are not mere formalities but acts of genuine appreciation and spiritual bonding. By expressing gratitude for the day's blessings, Manipuri communities reinforce their reliance on and respect for nature's gifts. Gratitude practices foster a mindset of sustainable living, where resource use is guided by principles of respect and responsibility rather than exploitation. The communal aspect of these practices also strengthens social cohesion, encouraging collective action toward ecological preservation. This shared sense of gratitude becomes a powerful motivator for environmental care and a buffer against the disconnection that often accompanies modern, industrialized societies.

Rituals for Agricultural Sustainability For the Manipuri, agriculture is not merely an economic activity but a spiritual endeavor intertwined with the rhythms of nature. Rituals performed at the start and end of farming cycles exemplify this deep connection. A farmer shared: "They dig in the corner of their field and give some fruits… to remember and pray that they don't do any harm to us or bring any curse to us." These offerings to land spirits symbolize a request for permission and protection, ensuring that agricultural activities remain harmonious with the natural world. By acknowledging the spiritual essence of the land, farmers foster a reciprocal relationship, where the land is treated as a living entity deserving of care and respect.

Such rituals also serve practical purposes, promoting agricultural sustainability. The acknowledgment of land spirits reinforces the community's commitment to avoiding harmful practices, such as over-cultivation or deforestation, which could disrupt the ecological balance. These spiritual practices act as a moral compass, guiding the community toward environmentally conscious farming methods that align with principles of climate resilience.

Climate Resilience Through Land-Based Spirituality The Manipuri Indigenous communities' spiritual practices offer valuable insights into climate resilience. By intertwining their cultural and spiritual identity with ecological stewardship, these practices address both immediate and long-term environmental challenges. The following are key ways in which land-based spirituality contributes to climate resilience: *Promoting sustainable practices*: The rituals and gratitude practices encourage mindful use of natural resources, reducing ecological degradation and fostering sustainability. *Strengthening social cohesion*: The communal nature of spiritual practices enhances solidarity, enabling collective efforts to address climate impacts. *Enhancing psychological resilience*: Spiritual rituals provide a sense of purpose and hope, mitigating the mental stress associated with climate uncertainty. *Preserving traditional knowledge*: Rituals serve as repositories of Indigenous ecological wisdom, ensuring its transmission to future generations.

The sacred connection to land and the spiritual practices of the Manipuri Indigenous communities underscore the importance of integrating cultural and spiritual dimensions into climate resilience strategies. By honoring the land through rituals, gratitude, and sustainable agricultural practices, these communities exemplify a holistic approach to living in harmony with nature. Their practices not only sustain their cultural heritage but also offer a blueprint for fostering ecological balance and resilience in the face of a changing climate. Recognizing and supporting such Indigenous worldviews is essential for creating inclusive and effective responses to global environmental challenges.

Observations of Climate Change and Environmental Shifts The Manipuri Indigenous communities' intimate connection with their environment enables them to keenly observe and interpret changes in climatic patterns and ecological systems. These observations, often framed within their spiritual worldview, provide crucial insights into the impacts of climate change.

Decreasing Rainfall and Water Scarcity Manipuri communities have noted significant shifts in rainfall patterns, which disrupt traditional water cycles and exacerbate water scarcity. One participant remarked: "The rainfall of Bangladesh over the last 10 years has been decreasing day by day… the water level is decreasing day by day." This decline in rainfall has profound implications for agriculture, reducing productivity and destabilizing the ecological balance. By recognizing these changes, the community adapts through water conservation practices and spiritual rituals aimed at invoking rainfall and maintaining harmony with nature's cycles. These adaptations demonstrate the role of Indigenous knowledge in responding to climate-induced challenges.

Loss of Biodiversity The loss of biodiversity is another alarming consequence of climate change observed by the Manipuri communities. A community member shared: "We are losing some species from our nature… these plants cannot survive in the present situation." Indigenous flora and fauna, integral to traditional practices and spiritual rituals, are disappearing, eroding both ecological and cultural resilience. This loss underscores the importance of preserving biodiversity not only for environmental stability but also for sustaining the cultural identity and spiritual practices of the community. Efforts to document and protect endangered species reflect a commitment to combating these challenges.

Heatwaves and Changing Seasons Extreme temperatures and erratic seasonal changes have further strained the community's adaptive capacity. One elder observed: "The heat wave of this year may have broken the record of the past 35–40 years." Such climatic extremes disrupt agricultural systems and pose health risks, especially to vulnerable populations. The community's spiritual framework, which includes rituals and prayers for climate stability, provides psychological resilience and fosters adaptive strategies to cope with these adversities. These practices demonstrate a deep understanding of environmental shifts and their impacts on daily life.

Decline of Indigenous Knowledge Systems As the Manipuri Indigenous communities face the pressures of modernization and environmental change, their traditional knowledge systems are at risk of erosion. This decline threatens the sustainability of their land-based practices and cultural identity.

Loss of Traditional Practices Traditional medicinal and culinary practices, rooted in Indigenous knowledge, are increasingly disappearing. One participant lamented: "People mix the Jenom leaves with some other ingredients… these practices are missing nowadays." Generational shifts and the allure of modernity have contributed to the abandonment of these time-honored traditions. Preserving such practices is crucial for maintaining the community's resilience and ecological harmony. Efforts to document and revitalize these traditions are vital steps toward sustaining their cultural heritage.

Intergenerational Knowledge Gaps The transmission of Indigenous knowledge from elders to younger generations is dwindling, leading to significant intergenerational gaps. An elder expressed concern: "This Indigenous knowledge will be abolished after a period of time… I can't transfer this knowledge." This gap poses a serious challenge to the continuity of traditional practices and spiritual wisdom. Initiatives to engage youth in learning and applying Indigenous knowledge are essential for bridging this divide and ensuring the community's adaptive capacity remains robust.

Dependence on Modern Alternatives The increasing reliance on modern agricultural and industrial systems has overshadowed traditional, sustainable methods. A participant observed: "Nowadays we are dependent on corporate houses… the technique of seed selection is forgotten." This dependence undermines the community's self-sufficiency and ecological harmony. Reviving traditional practices, such as seed selection and organic farming, can reduce reliance on external systems and restore sustainable livelihoods.

The sacred connection to land and the spiritual practices of the Manipuri Indigenous communities underscore the importance of integrating cultural and spiritual dimensions into climate resilience strategies. By honoring the land through rituals, gratitude, and sustainable agricultural practices, these communities exemplify a holistic approach to living in harmony with nature. Their practices not only sustain their cultural heritage but also offer a blueprint for fostering ecological balance and resilience in the face of a changing climate. Recognizing and supporting such Indigenous worldviews is essential for creating inclusive and effective responses to global environmental challenges.

Architectural and Cultural Sustainability Sustainability for Manipuri Indigenous communities extends beyond spiritual practices to their architectural and cultural systems. By embedding traditional knowledge into housing, infrastructure, and rituals, these communities create structures that are both environmentally and culturally sustainable.

Traditional Housing Techniques Manipuri traditional housing exemplifies environmental sustainability through the use of natural materials and age-old techniques. A community elder noted: "Our housing system… was made of bamboo… it has a longevity of 80–100 years." These houses are built with renewable materials like bamboo and thatch, which are abundant, biodegradable, and culturally significant. The construction methods ensure durability and harmony with the local climate, reducing the environmental footprint. Moreover, the building process itself is imbued with rituals that pay respect to nature, reinforcing the community's spiritual connection to the land. By retaining these techniques, Manipuri communities preserve their cultural heritage and promote sustainable living. Traditional housing serves as a model for integrating environmental consciousness into modern architectural practices.

Cultural Symbolism in Indigenous-Led Infrastructures Symbolism plays a crucial role in Manipuri infrastructure, aligning construction practices with spiritual and natural principles. A participant described: "The counting must end with water… because water is a symbol of coldness." This ritual reflects the deep spiritual significance of water as a source of life and balance. Such practices ensure that construction aligns with cultural values and ecological harmony. Buildings are not mere physical structures; they are embodiments of cultural narratives, preserving the community's identity and fostering a sense of belonging. By integrating cultural symbolism into infrastructure, the Manipuri communities demonstrate how traditional wisdom can guide sustainable development. These practices inspire broader applications of Indigenous knowledge in modern urban planning and design.

Rituals for Building Harmony The process of construction in Manipuri culture is accompanied by rituals that honor the creatures displaced by human activity. One elder shared: "There are different worships… to pray to the insects that used to live here before us." These rituals acknowledge the impact of construction on local ecosystems and seek to restore spiritual balance. By performing these ceremonies, the community ensures that their actions respect and coexist with the natural world. These rituals also strengthen community cohesion, as the collective act of honoring nature reinforces shared values and responsibilities. Such practices highlight the importance of ethical and sustainable approaches to development, offering a blueprint for ecological and spiritual balance.

Strategies for Cultural and Climate Adaptation Manipuri Indigenous communities are actively developing strategies to adapt to climate challenges while preserving their cultural heritage. These efforts encompass innovative approaches to education, documentation, and the creation of model villages that integrate traditional and modern practices.

Model Indigenous Villages The concept of model Indigenous villages represents a holistic approach to cultural and environmental sustainability. A participant envisioned: "There will be Manipuri household structure and other cultural settings… to preserve these." These villages aim to recreate traditional living environments, showcasing Manipuri architecture, cultural practices, and ecological principles. By integrating modern amenities with traditional designs, these villages serve as living laboratories for sustainable development and cultural preservation. Model villages also act as centers for community learning and tourism, creating opportunities for economic development while fostering pride in Indigenous heritage. This innovative approach demonstrates the potential for Indigenous knowledge systems to shape resilient and inclusive communities.

Documentation of Indigenous Knowledge Preserving Indigenous knowledge is essential for sustaining cultural and ecological resilience. A community member emphasized: "If any initiative is taken to document this knowledge it will be preserved… for 100 or 200 years." Documentation initiatives involve recording oral histories, rituals, and traditional practices to ensure their transmission to future generations. These efforts address the risk of knowledge loss due to generational shifts and modern influences. By creating accessible repositories of Indigenous knowledge, the Manipuri communities safeguard their cultural legacy and provide valuable resources for academic research and policy-making. Documentation also enhances cross-cultural understanding, showcasing the relevance of Indigenous perspectives in addressing global challenges.

Integrating Education with Indigenous Values Education serves as a powerful tool for bridging the gap between traditional knowledge and contemporary challenges. One participant advocated: "We must bring those perspectives back… for our educational curriculum policy-making." Incorporating Indigenous practices and values into educational curricula fosters cultural pride and environmental stewardship. This approach equips younger generations with the tools to navigate modern challenges while staying rooted in their heritage. Educational programs that integrate Indigenous perspectives also promote cross-disciplinary learning, combining ecological science with cultural wisdom. These initiatives demonstrate the potential for education to serve as a catalyst for sustainable development and social cohesion.

The Manipuri Indigenous communities' strategies for architectural and cultural sustainability highlight the intricate connections between traditional knowledge, spiritual practices, and ecological resilience. By honoring their heritage through sustainable housing, symbolic rituals, and innovative adaptation strategies, these communities offer a model for integrating cultural identity with climate resilience. Through the preservation of Indigenous knowledge, the creation of model villages, and the integration of traditional values into education, the Manipuri communities reaffirm their commitment to cultural and environmental sustainability. These efforts not only strengthen their resilience but also provide valuable insights for global efforts to address climate change and cultural preservation. Recognizing and supporting such Indigenous-led initiatives is essential for fostering a more inclusive and sustainable future.

Discussion: Land-Based Spirituality as a Framework for Climate Resilience

The findings highlight the unique and transformative role of land-based spirituality in fostering climate resilience among Manipuri Indigenous communities. By intertwining spiritual, cultural, and ecological practices, these communities provide a profound counterpoint to the shortcomings of mainstream Western environmental management. This discussion contextualizes these findings within broader environmental, cultural, and policy frameworks while drawing on existing literature to emphasize the global relevance of Indigenous perspectives on climate adaptation and sustainability.

Bridging the Disconnect Western Environmental Management vs. Indigenous Practices. Mainstream Western environmental management often operates within a utilitarian framework that commodifies nature, prioritizing economic development over ecological harmony. This approach has led to numerous challenges, including biodiversity loss, soil degradation, and climate instability (Robbins, 2012; Wilson, 2016). Manipuri Indigenous communities have witnessed firsthand the detrimental impacts of these practices, particularly the erosion of traditional flora critical to their cultural and ecological systems (Datta et al., 2024). In contrast, land-based spirituality emphasizes a reciprocal relationship with the environment, fostering practices that sustain ecological balance. Rituals such as the *Umang Lai Haraoba* symbolize the sacred bond between humans and nature, challenging the anthropocentric ethos of Western environmental management (Chongtham et al., 2018; Kimmerer, 2013). By integrating spiritual accountability with ecological stewardship, Manipuri practices provide a holistic framework for addressing environmental challenges.

Rituals as Catalysts for Sustainable Practices Manipuri farmers view agriculture as a spiritual endeavor deeply intertwined with the rhythms of nature. Rituals performed at the start and end of farming cycles serve dual purposes: they honor land spirits and promote sustainable farming methods. Manipuri communities perform rituals that seek forgiveness from non-human beings displaced beings and express gratitude to land spirits. One elder explained: "We must ask for forgiveness from them [non-human being] as we have destroyed their house to build ours… This puja helped me to create respect for other living things."These practices align with principles of regenerative agriculture by enhancing soil fertility and minimizing dependency on chemical inputs (Sharma et al., 2017). As observed in other Indigenous communities, such approaches significantly contribute to biodiversity conservation and climate adaptation (Altieri, 2004; Tilman et al., 2002).

Gratitude and Resilience Gratitude rituals, such as evening prayers, underscore the interconnectedness of life and resource use. These daily practices instill a mindset of respect and responsibility, contrasting starkly with exploitative practices rooted in Western systems (Cajete, 2000). Beyond fostering environmental care, these rituals enhance social cohesion, creating a collective framework for addressing ecological challenges (Berkes, 2008).

Addressing Climate Impacts Through Spiritual Worldviews

Manipuri communities possess an acute awareness of environmental changes, interpreting shifts in rainfall patterns, biodiversity loss, and temperature extremes through their spiritual lens. This perspective provides invaluable data for understanding localized climate impacts (Khumbongmayum et al., 2005). For instance, their observations of declining water levels have prompted adaptations, including water conservation practices and rituals invoking rainfall (Bhadra et al., 2021).

Biodiversity as Cultural and Ecological Capital The loss of Indigenous flora and fauna threatens both ecological stability and cultural identity. Manipuri sacred groves, which serve as biodiversity refuges, exemplify the community's commitment to preserving ecological and cultural resilience (Chapin et al., 2009). These groves align with global conservation goals, offering a model for integrating traditional knowledge into contemporary environmental governance (Gadgil et al., 1993).

Revitalizing Indigenous Knowledge Systems The erosion of traditional practices and intergenerational knowledge gaps pose significant threats to the sustainability of Manipuri land-based spirituality. Elders lament the decline of traditional medicinal and culinary practices, underscoring the need for documentation and revitalization efforts (Smith, 1999; Mohanty, 2003). Programs aimed at engaging youth in learning these practices are essential for ensuring continuity and fostering a renewed sense of cultural pride (Whyte, 2018).

Balancing Tradition and Modernity Dependence on modern agricultural systems has overshadowed traditional, sustainable methods, leading to a loss of self-sufficiency (Sharma et al., 2017). However, integrating Indigenous practices with modern innovations presents an opportunity to create hybrid systems that leverage the strengths of both approaches. For example, combining traditional seed selection techniques with modern agricultural science could enhance resilience and reduce dependency on external resources (Singh et al., 2020).

Traditional Housing as a Model for Resilience Manipuri traditional housing techniques, which use renewable materials like bamboo, exemplify environmental sustainability. These structures are not only durable but also harmonized with the local climate, reducing the community's carbon footprint (Chongtham et al., 2018). By embedding spiritual and cultural values into construction practices, such as rituals honoring displaced creatures, these homes serve as symbols of ecological and spiritual balance (Datta et al., 2024). The use of cultural symbolism in Manipuri infrastructure, such as ending construction rituals with water as a symbol of life, reinforces the community's spiritual and ecological values. These practices highlight the potential of traditional knowledge to inform modern sustainable design, inspiring global urban planning frameworks to adopt more ethical and inclusive approaches (Kothari et al., 2013).

Lessons for Global Environmental Governance The Manipuri emphasis on reciprocity, gratitude, and sustainability offers valuable lessons for addressing global environmental challenges. Studies show that Indigenous-managed territories often outperform protected areas in biodiversity conservation, underscoring the efficacy of Indigenous stewardship (Chapin et al., 2009). Recognizing and integrating such knowledge into global frameworks is crucial for achieving climate justice and ecological balance (Wilson, 2016).

Incorporating Indigenous perspectives into environmental education fosters cross-cultural understanding and respect for diverse worldviews. Manipuri practices of gratitude and ecological stewardship can inspire ethical shifts in global environmental movements (Cajete, 2000; Smith, 1999). Advocacy for Indigenous rights and knowledge systems must remain central to these efforts, ensuring that Indigenous voices are amplified in policy-making and governance (Whyte, 2018).

The findings from Manipuri Indigenous communities emphasize the transformative potential of land-based spirituality in addressing environmental challenges. By prioritizing harmony with nature, these communities offer a compelling alternative to the exploitative practices of mainstream Western environmental management. Their spiritual practices not only sustain cultural heritage but also provide practical strategies for fostering climate resilience.

Recognizing and supporting such Indigenous perspectives is essential for creating inclusive and effective responses to global environmental challenges. As the global community grapples with the impacts of climate change, integrating Indigenous knowledge and spirituality into environmental governance offers a path toward a more equitable and sustainable future.

Chapter 9
Challenging Human-Created Climate Crisis and Land-Based Spirituality as Climate Action from Coastal Area of Rakhine Indigenous Community Perspectives

This chapter explores the human-created climate crisis from the perspective of the Rakhine Indigenous community, emphasizing their land-based spirituality as a critical lens for understanding and responding to environmental challenges. Situated in a coastal region highly vulnerable to rising sea levels and extreme weather events, the Rakhine community navigates the intersections of ecological degradation, spiritual ethics, and resilience. The chapter delineates two overarching themes: the anthropogenic causes of climate crises and land-based spirituality as climate action. Subthemes include the impacts of human greed, deforestation, and industrial waste; the intrinsic value of land as a naturally created entity; and the role of spiritual principles such as Panchashila in fostering ecological stewardship. The findings highlight the community's reliance on practices like afforestation, water conservation, and the symbolic reverence for trees and water, which are deeply rooted in their spiritual worldview. By integrating traditional ecological knowledge (IEK) with spirituality, the Rakhine community demonstrates a holistic approach to resilience, balancing cultural heritage with adaptive practices. The chapter concludes with actionable insights into community-based adaptation strategies, the challenges of mainstream marginalization, and the necessity for inclusive policy frameworks that center Indigenous voices in climate action.

The anthropogenic causes of climate change have been extensively documented. Industrialization, deforestation, and greenhouse gas emissions are among the primary drivers, leading to rising global temperatures, melting ice caps, and extreme weather events (IPCC, 2021). Coastal areas, like those inhabited by the Rakhine Indigenous community, are particularly vulnerable due to rising sea levels and increased cyclone activity (Rahman et al., 2020). For instance, Bangladesh's coastal belt has experienced severe cyclones such as SIDR (2007) and AILA (2009), causing extensive damage to life and property (Mallick & Rahman, 2017). These events disproportionately affect marginalized communities, exacerbating socio-economic vulnerabilities (Islam & Winkel, 2017). The role of human activity in exacerbating natural disasters is a recurring theme in the literature. Greed-driven deforestation,

R. Datta, *Decolonizing Climate Adaptation*,
https://doi.org/10.1007/978-981-96-9140-1_9

unregulated industrial waste disposal, and unsustainable agriculture contribute to environmental degradation (Shiva, 2016). In particular, the destruction of mangrove forests, which act as natural buffers against cyclones, has left coastal communities increasingly exposed (Ghosh et al., 2018). The urgency of mitigating these human-induced impacts is evident, but solutions often overlook the localized knowledge and practices of Indigenous communities.

Indigenous ecological knowledge (IEK) offers a valuable lens for addressing climate challenges. Rooted in centuries of interaction with local ecosystems, IEK emphasizes sustainable resource management and the interconnectedness of all life forms (Berkes, 2018). In the context of coastal communities, IEK includes practices such as afforestation, water conservation, and disaster preparedness (McGregor, 2014). For example, the use of traditional cyclone shelters and food preservation techniques in Bangladesh demonstrates the effectiveness of Indigenous strategies (Rahman et al., 2020). However, the marginalization of Indigenous voices in policy-making has limited the integration of IEK into broader climate action frameworks (Whyte, 2017). Studies indicate that incorporating Indigenous perspectives can enhance the resilience of climate adaptation strategies (Kimmerer, 2013). In the case of the Rakhine community, their reliance on natural elements like trees and water aligns with global calls for nature-based solutions (Chausson et al., 2020).

Land-based spirituality forms a core component of many Indigenous worldviews, linking ethical principles with environmental stewardship (Cajete, 2000). Among Buddhist Indigenous communities, principles such as non-violence (Ahimsa) and compassion extend to the natural world (Lama, 2020). The Rakhine's adherence to Panchashila—a code of ethical living—reflects this integration of spirituality and ecology. Scholars argue that spirituality can drive pro-environmental behavior by fostering a sense of interconnectedness and moral responsibility (Rasmussen et al., 2021). For instance, the practice of showing respect to trees and water, as described by the Rakhine, aligns with global movements for ecological reverence (Taylor, 2010). Similarly, rituals centered around natural elements can reinforce community cohesion and collective action for conservation (King, 2013).

Despite its potential, the role of land-based spirituality in climate action remains underexplored in academic discourse. Studies often prioritize technological and policy solutions, neglecting the cultural and spiritual dimensions of environmentalism (Verlie, 2021). This gap highlights the need for interdisciplinary approaches that value Indigenous spiritual frameworks as integral to sustainability.

Coastal regions face unique challenges in implementing climate action. Strategies such as mangrove restoration, community-based disaster preparedness, and livelihood diversification have shown promise (Adger et al., 2005). For example, mangrove afforestation projects in the Sundarbans have reduced cyclone impacts and enhanced biodiversity (Rahman et al., 2020). However, these initiatives often encounter barriers such as inadequate funding, policy misalignment, and lack of community engagement (Haque et al., 2019). The participation of Indigenous communities in climate action is critical. Studies emphasize the importance of co-creating solutions with local stakeholders to ensure cultural relevance and effectiveness (Reed et al., 2018). In the context of the Rakhine community, integrating their land-based spirituality with scientific approaches could yield innovative strategies for coastal resilience (Berkes, 2018).

A significant gap in addressing climate challenges lies in the undervaluation of Indigenous knowledge. Policymakers and researchers must prioritize the documentation and integration of practices such as those employed by the Rakhine. For instance, their emphasis on tree planting—particularly species like banyan and tamarisk—addresses both ecological and cultural needs. Collaborative efforts are needed to mainstream such practices into national and global climate strategies. The preservation of land-based spirituality is essential for sustaining Indigenous resilience. The Rakhine community's reliance on spiritual principles such as Panchashila underscores the need for policies that protect cultural heritage. Initiatives like incorporating Indigenous spirituality into education curricula and supporting community-led cultural programs can strengthen these frameworks.

Scholars like Whyte (2017) argue that cultural erosion undermines community resilience, making spiritual preservation a critical component of climate action. Ensuring that younger generations inherit and practice these traditions is particularly important, as highlighted by the Rakhine's efforts to teach children about spirituality and sustainability.

Community-based adaptation (CBA) approaches must be scaled up to address the specific vulnerabilities of coastal regions. The Rakhine's experiences with cyclones and rising sea levels illustrate the importance of localized solutions. Key actions include:

- Cyclone preparedness: Building and maintaining cyclone shelters that accommodate cultural and spiritual needs, as advocated by Haque et al. (2019).
- Sustainable livelihoods: Providing training in traditional and modern skills, such as knitting, sewing, and sustainable agriculture, to enhance economic resilience.
- Reforestation: Supporting community-led tree planting initiatives, with a focus on species that mitigate disaster impacts and align with spiritual values.

The disconnect between policy frameworks and grassroots realities remains a critical challenge. Policymakers must engage with Indigenous communities to co-create solutions that reflect local priorities. For the Rakhine, this includes addressing land tenure issues, ensuring equitable access to resources, and providing financial support for conservation efforts. Berkes (2018) emphasizes the need for adaptive governance structures that integrate scientific and Indigenous knowledge systems. Such frameworks can enhance the scalability and sustainability of climate action in coastal regions. The underrepresentation of land-based spirituality in climate research highlights the need for interdisciplinary studies that bridge environmental science, anthropology, and religious studies. Exploring the Rakhine's spiritual practices through this lens can uncover novel insights into the relationship between spirituality and ecological resilience.

Rasmussen et al. (2021) suggest that incorporating cultural and spiritual dimensions into climate discourse can inspire more holistic and inclusive approaches. Future research should prioritize participatory methodologies to amplify Indigenous voices and co-create knowledge.

The human-created climate crisis poses significant challenges for coastal communities like the Rakhine. However, their land-based spirituality offers a powerful framework for resilience and adaptation. By recognizing the value of Indigenous ecological knowledge, supporting cultural preservation, and bridging policy gaps, we can foster more effective and inclusive climate action. Interdisciplinary research and community-based adaptation initiatives are essential for addressing the critical needs identified in this review. Embracing these approaches can ensure that Indigenous perspectives, such as those of the Rakhine, play a central role in shaping sustainable futures.

Research Findings

The finding section focuses on the environmental challenges and spiritual resilience of the Rakhine people. The primary theme of this discourse is the human-created climate crisis, which manifests through deforestation, industrial pollution, and the exploitation of natural resources, leading to rising sea levels, frequent cyclones, and loss of biodiversity. Subthemes under this include the ethical critique of human greed, the impacts of environmental degradation on vulnerable coastal areas, and the need for immediate mitigation efforts. Complementing this is the theme of land-based spirituality as a transformative approach to climate action. Subthemes include the integration of Panchashila of non-violence, truthfulness, and respect for nature\u2014into daily practices, the symbolic and practical importance of conserving water and trees as sacred elements, and the transmission of these values to younger generations through education and community engagement. By merging traditional spiritual ethics with actionable environmental practices, the Rakhine community offers a holistic model of resilience and adaptation to confront the pressing climate crisis (Table 9.1).

Land and Human Relationship The relationship between Land and humanity is profound and multifaceted. This section explores three key dimensions of this relationship: the definition of nature, human dependency on nature, and human impact on nature, as articulated by the Rakhain Indigenous community. Their perspectives offer invaluable insights into understanding and addressing environmental challenges through a spiritual and land-based lens.

Meanings of Land The Rakhain Indigenous community conceptualizes Land as "naturally created creation," emphasizing its existence independent of human intervention. They view Land as a dynamic and eternal entity comprising forests, oceans, mountains, and air, which cannot be created by human beings. One respondent eloquently stated, "Land is naturally created creation, and which are not created by human being. For example—forests, oceans, mountains. Air; we cannot see but we do feel. They were long before four kalpas, they are still here, and they will be the world in the future." This definition underscores a deep respect for the intrinsic and autonomous value of nature, rejecting the anthropocentric notion of humans as the

Table 9.1 Shows the challenging human-created climate crisis and land-based spirituality as climate action from coastal area of rakhine indigenous community perspectives

Main themes	Subthemes
Human-created climate crisis	– Impacts of deforestation, industrial waste, and exploitation of resources
	– Loss of biodiversity and rising sea levels in coastal regions
	– Greed and unchecked human desires as ethical critiques
Land-based spirituality	– Panchashila as a moral framework: non-violence, truthfulness, respect for nature
	– Reverence for trees and water: symbolic and practical significance
	– Transmitting ecological values through education and community practices
Human-land relationships	– Meanings of land: a naturally created, autonomous entity
	– Human dependency on land: trees as protectors, providers, and disaster mitigators
	– Human impact: pollution, deforestation, and consequences of neglect
Resilience and adaptation strategies	– Cyclone preparedness: shelters and stored resources
	– Community-led tree planting: Banyan and tamarisk for resilience
	– Livelihood diversification: integrating traditional and modern skills
Cultural identity and preservation	– Challenges from mainstream aggression and systemic marginalization
	– Educational initiatives for language and spiritual knowledge transmission
	– Role of spirituality in fostering community cohesion and ethical living

ultimate creators or controllers. The community's belief in the enduring presence of Land over kalpas—long periods of time in Buddhist cosmology—reflects their philosophical and temporal understanding of the natural world. It positions Land as a constant, transcending human life and activities. This view fosters a mindset of stewardship, recognizing the intergenerational responsibility to sustain and honor the natural environment.

The Rakhain people emphasize the symbiotic relationship between Land and human survival. They assert, "The world sustains on the nature. For example—trees, air, etc. The more trees we will have, the easier it will be for breathing. As we live in the coastal area, trees will save us from heavy rain, flooding, and this will be good for us." Here, the emphasis lies on nature's life-supporting functions, particularly in the context of the coastal ecosystem. Trees, for instance, play a pivotal role in regulating climate, preventing soil erosion, and protecting communities from natural disasters. This holistic meaning of Land also carries a cautionary note about the consequences of neglecting its conservation. As one respondent warned, "If we don't conserve the nature, we will not be able to survive on this land. We need to know how we can survive the natural disasters like flood, earthquake, drought, heavy rain, etc." Such statements reflect the community's acute awareness of the interconnectedness between ecological preservation and human resilience, particularly in disaster-prone regions.

Human Dependency on Land The Rakhain Indigenous community's insights reveal a profound dependency on Land for survival and well-being. Their reliance extends beyond material sustenance to include spiritual and cultural dimensions. They highlight the essential services provided by trees, stating, "Trees will stop soil erosion, give shades from sun, bring rain. The more trees we will have, the easier it will be for breathing. As we live in the coastal area, trees will save us from heavy rain, flooding, and this will be good for us." Trees are seen as protectors and providers, integral to the community's physical safety and environmental balance. Moreover, nature's contributions to biodiversity and ecological harmony are deeply valued. As another respondent noted, "The contents of the Land help every living being. Some trees give medicine; some are food of other animals. The Land is helping every creature to survive." This recognition of nature's generosity aligns with traditional ecological knowledge, emphasizing interdependence among all living beings. By viewing Land as a caregiver, the Rakhain people advocate for a reciprocal relationship where humans must respect and nurture the environment. The dependency on Land is also framed within a broader existential context. One participant reflected, "When the earth cannot bear anymore, earthquakes happen. Sea level is rising. So, we need to learn the Land and how to live with and respect it." This statement underscores the consequences of exceeding nature's carrying capacity, highlighting the urgency for sustainable living practices. The rising sea levels and increasing frequency of earthquakes are seen as manifestations of nature's limits being tested, necessitating immediate action to restore balance.

Human Impact on Land The Rakhain community's reflections on human impact reveal a deep concern for the detrimental effects of human activities on the environment. They identify greed and unchecked desires as the root causes of environmental degradation. As one respondent lamented, "Human being is the most dangerous creature. They have no end of their desires. The more they get, the more they want. An animal like a lion hunts for the sake of his hunger. And when his stomach is full, he leaves the remaining food for other animals. But humans just want everything." This critique of human behavior contrasts sharply with the natural restraint observed in the animal kingdom, highlighting the moral and ethical dimensions of environmental stewardship.

The destructive consequences of human activities are vividly described. For example, "Currently, we can see that trees are being cut, and the atmosphere suffers from this. Thermal power plantations have bad impacts on the climate. Wastages from different industries are coming out to air and water." This observation points to specific actions—deforestation, industrial pollution, and unsustainable energy production—that harm the environment. The community's firsthand experience of these impacts lends credibility to their warnings and calls for corrective measures. The ultimate cost of human exploitation of Land is expressed in stark terms: "We are destroying the Land for our greed, and someday the world will come to an end because of this greed. No animal can do harm to nature, but humans do." This prophetic statement captures the gravity of the ecological crisis and serves as a moral indictment of humanity's role in accelerating environmental collapse. It also highlights the uniqueness of human agency, capable of both creating and resolving crises, urging a shift toward responsible and compassionate interactions with the natural world.

Broader Implications The findings from the Rakhain Indigenous community offer significant implications for environmental sustainability, education, and policy-making. By framing Land as an autonomous and interdependent entity, their perspectives challenge anthropocentric paradigms that dominate modern development narratives. Instead, they advocate for a relational approach that prioritizes harmony, respect, and reciprocity. The emphasis on human dependency underscores the need to integrate traditional ecological knowledge into contemporary practices. For coastal communities like the Rakhain, nature-based solutions—such as afforestation, biodiversity conservation, and sustainable resource management—are essential for mitigating the impacts of climate change and natural disasters. These insights can inform localized adaptation strategies and strengthen community resilience. The critique of human impact calls for a reevaluation of economic and industrial systems that prioritize profit over environmental health. Policies promoting renewable energy, circular economies, and pollution control can align with the Rakhain community's vision of ecological balance. Additionally, fostering ethical and spiritual values in education can inspire individuals to adopt sustainable lifestyles and advocate for systemic change.

The Rakhain Indigenous community's understanding of Land and its relationship with humanity provides a powerful framework for addressing environmental challenges. By defining Land as an eternal and autonomous entity, emphasizing human dependency, and critiquing human impact, they offer a holistic perspective rooted in spirituality and lived experience. Their insights highlight the urgency of adopting sustainable practices, respecting ecological limits, and fostering a harmonious coexistence with the natural world. This relational approach not only enriches environmental discourse but also serves as a call to action for collective responsibility and stewardship.

Land-Based Spirituality and Land Conservation The intersection of spirituality and environmental stewardship is central to the perspectives of the Rakhain Indigenous community. This section delves into their philosophy and ethics, the role of spirituality in conservation, and their efforts in teaching younger generations about the importance of nature.

Land-Based Spirituality Philosophy and Ethics The Rakhain community's approach to spirituality emphasizes ethical living and ecological responsibility. Central to their philosophy is Panchashila, a set of moral precepts forming the foundation of their way of life. A participant articulated, "In my spirituality, Panchashila has been defined: do not kill, do not lie, do not steal, non-aggression, and not being intoxicated. Buddha has said whoever maintains the Panchashila is a Buddha. People do not become Buddha by birth. If a person is born in a Buddhist family but does not maintain Panchashila, he is not a Buddhist." This statement underscores the active and intentional practice required to embody spiritual principles, emphasizing actions over heritage.

The community's ethical framework extends to environmental conservation. They view water and trees not as objects of worship but as entities deserving respect. "Buddha told to save water. It's not praying but to show respect to water. If there is no respect, there is no love, and people easily can do harm to nature. In the same way, Buddha has told to save the Bodhi tree, not to pray the tree," explained one respondent. Such perspectives advocate for reverence toward natural elements, fostering a culture of preservation rather than exploitation. The overarching goal of their spiritual practice is peace—both within individuals and the broader environment. One participant stated, "Religion is for peace. But if there is no peace in the religion, what will we do with the religion? Peace is the main theme in every religion. But we have spoiled the peace." This critique points to the importance of aligning spiritual values with actions, particularly in addressing environmental challenges.

Role of Spirituality in Conservation The Rakhain community's spirituality plays a pivotal role in their environmental conservation efforts. Their teachings and practices underscore the interconnectedness of all life and the responsibility to protect natural resources. One respondent recounted, "Buddha became a Buddha under a Bodhi tree and stayed seven more days to show the importance of nature. The tree and Land have given you shelter; you have to save them. Try to plant trees so that deforestation doesn't happen." This narrative not only highlights the symbolic significance of trees in their spiritual tradition but also advocates for practical actions such as reforestation. Water conservation is another critical aspect of their spirituality. Participants emphasized the need to prevent water pollution and promote sustainable practices. "The water we use should not be polluted. Buddha told to save water and not harm it. Some people worship water. It is not a prayer but a form of showing respect to water," explained a community member. Such teachings encourage a shift from exploitative to respectful interactions with natural resources.

Their approach to conservation is deeply rooted in love and respect for all living beings. As one respondent observed, "By living with love and respect toward every living being, we can save the nature. We need to accept the good things and leave the bad things for the world to be better." This principle advocates for a harmonious coexistence, emphasizing that ethical behavior toward others—both human and non-human—is essential for ecological balance.

Teaching Young Generations The Rakhain community recognizes the importance of imparting spiritual and ecological knowledge to younger generations to ensure the continuity of their values and practices. They stress the need for self-reliance and sustainable living. "If the generation can produce their own food and clothing, then the poverty will be reduced. Our youth need to learn skills like planting vegetables, knitting, sewing, and even computer knowledge. This is how they can sustain themselves and help the community," shared one respondent. This focus on practical skills reflects a holistic approach to education that integrates environmental, economic, and social dimensions. Efforts to teach children about their cultural and spiritual heritage are ongoing, albeit challenging. One participant noted, "I try in my temple to teach the children about the language and spirituality. But I cannot do it alone because I am occupied with many activities. My job is to teach spirituality, not language. A teacher is needed to teach the children." This statement highlights the resource constraints faced by the community in preserving their traditions and educating the next generation.

The community's teachings also emphasize disaster preparedness and resilience, linking spiritual principles to practical actions. "We try to make the people understand that when there is a cyclone, keep your home safe and come to the cyclone center. Money will come and go, but if you do not live, what will you do with money and property?" explained a participant. Such guidance reflects the integration of spirituality with pragmatic measures to address environmental challenges.

Human-Created Disasters and Resilience The Rakhain Indigenous community's lived experiences and responses to human-created disasters offer valuable lessons in resilience and adaptation. This section explores the impact of human-created disasters, disaster preparedness, and the roles of community and government in mitigating their effects.

Impact of Human-Created Disasters The community has faced numerous natural disasters, including the devastating cyclone SIDR in 2007. One participant recounted, "In 2007, we have seen SIDR. The roof of our house was blown away, and the base of the house was broken. The dam was damaged, and the water entered the village." Such firsthand accounts highlight the physical and emotional toll of these events, underscoring the vulnerability of coastal communities. The aftermath of SIDR revealed the severe challenges of survival. "During SIDR, we starved for three days as there was no place for cooking. We suffered and learned techniques on how to survive better during disasters," shared a respondent. This statement underscores the importance of experiential learning in building resilience and improving future responses to disasters. The psychological impact of recurring disasters is significant, creating a pervasive sense of insecurity. "Cyclones create panic among us because we often don't know where to go. Many are left without shelter or adequate supplies, making survival extremely difficult," explained a participant. These experiences highlight the need for robust support systems to alleviate the fear and uncertainty associated with natural disasters.

Disaster Preparedness The Rakhain community has developed various strategies to prepare for and respond to natural disasters. Their approach emphasizes self-reliance and community solidarity. "We asked the people of my community to store dry food, drinking water so that they have something to eat when there is no situation for cooking," shared a participant. Such practices reflect a proactive mindset, aiming to mitigate the immediate impacts of disasters. Cyclone centers play a crucial role in ensuring safety during emergencies. "We maintain cyclone centers and allow everyone to take shelter there. Before SIDR, people didn't come to the centers, but now they understand their importance," explained a community member. These centers serve as vital refuges, fostering a sense of security and community during crises.

Despite their efforts, the community acknowledges the limitations of their resources and the need for external support. "Cyclone centers must have proper provisions for safety and shelter. We rely on community efforts as government relief is often inadequate and mismanaged," observed a participant. This statement highlights the critical need for collaborative approaches to disaster preparedness and response.

Community and Government Roles The Rakhain community's experiences reveal significant gaps in government support and the challenges posed by corruption and inefficiency. "Political leaders take all the relief given by the government, and in the end, we common people get nothing. For example, government allocated 25 thousand crore taka for our area, but people got only 5000 taka," shared a participant. Such accounts underscore the systemic barriers to equitable disaster relief and recovery.

Improved planning and implementation are essential to address these challenges. "If there was proper planning and implementation, we could mitigate many disasters. The government does take action, but political corruption hinders the efforts," explained a respondent. This critique calls for greater accountability and transparency in disaster management.

The community also emphasizes the need for tailored support to address their unique vulnerabilities. "We need cyclone centers specifically for our community to ensure safety during disasters. Without proper support, we are left vulnerable," stated a participant. This perspective highlights the importance of inclusive and culturally sensitive approaches to disaster risk reduction.

The Rakhain Indigenous community's perspectives on land-based spirituality, Land conservation, and disaster resilience provide a holistic framework for addressing environmental and social challenges. Their spiritual principles and practical actions underscore the interconnectedness of ethical living, ecological stewardship, and community well-being. By integrating these insights into broader policy and education initiatives, we can foster a more sustainable and equitable future.

Cultural Identity and Spirituality Cultural identity and spirituality are vital elements of the Rakhain Indigenous community's worldview, offering both a foundation for resilience and a framework for addressing contemporary challenges. This section examines the preservation of Rakhain culture, challenges posed by mainstream aggression, and the role of spirituality in overcoming these challenges.

Preservation of Rakhain Culture The preservation of cultural heritage is a pressing concern for the Rakhain community, particularly as their language and traditions face the risk of extinction. One participant articulated this urgency: "Nobody ever thought about our language or teaches our children in Rakhain language. If our youngsters do not know about the language, this will be lost." The erosion of linguistic heritage not only jeopardizes cultural identity but also disconnects younger generations from their ancestral wisdom and spiritual practices. Efforts to preserve culture often rely on individual initiatives, which are resource-intensive and limited in scope. A respondent shared, "I try in my temple to teach the children about the language and spirituality. But I cannot do it alone because I am occupied with many activities. My job is to teach spirituality." This dual responsibility highlights the need for institutional support to ensure that cultural preservation extends beyond personal efforts.

The lack of formal recognition and resources for Rakhain cultural preservation is starkly evident during commemorative events. As one participant observed, "When the 21st February comes, so many people come to me for interviews. But nobody ever thought about our language and cultural preservation." This statement underscores the performative Land of such engagements, which fail to translate into actionable support for the community's cultural needs. Addressing this disconnect requires sustained engagement, inclusive policy-making, and educational reforms to integrate Rakhain cultural elements into broader societal narratives.

Challenges from Mainstream Aggression The Rakhain community's cultural identity and spiritual practices are further strained by systemic aggression and marginalization from the mainstream society. One participant described the prevailing attitude as follows: "Religion should be polite, humble, and peaceful. But the mainstream uses the 'Might is Right' approach. They think minorities are insignificant and neglect us." This critique highlights the disparity between the professed ideals of religious harmony and the lived reality of discrimination faced by minorities.

Systemic challenges manifest in various forms, including conversion attempts and social pressures. A respondent noted, "We are bound to obey their rules; they are creating all the problems, not us. Conversion attempts, social pressures, and systemic aggression weaken our community's identity." These practices not only threaten the community's autonomy but also erode the trust and cohesion necessary for cultural preservation.

The younger generation's limited knowledge of their spiritual heritage exacerbates the vulnerability to external influences. "The message of our spirituality should be sent to the young generation. Less knowledge of our own religion makes people susceptible to conversion to other religions," explained a participant. This observation underscores the critical need for educational initiatives that empower youth with a deeper understanding of their cultural and spiritual roots, fostering resilience against external pressures.

Role of Spirituality in Overcoming Challenges Despite the challenges, the Rakhain community's spirituality offers a robust framework for resilience and harmony. Central to their spiritual ethos is the principle of universal love and compassion. One participant reflected, "My spirituality has told me to love other living beings. The way I keep myself healthy and safe, I should treat other living beings like that, and that is my duty." This perspective encourages a relational approach to life, emphasizing interconnectedness and mutual care. The practice of Panchashila—a set of ethical precepts—further reinforces the community's commitment to peace and integrity. "By maintaining Panchashila, one becomes a Buddha, not by birth but by actions. We need to share this knowledge with the younger generation to preserve our traditions," shared a participant. This statement highlights the transformative potential of ethical living, positioning spirituality as a dynamic and inclusive practice.

Spirituality also serves as a guiding principle for community cohesion and self-reliance. "Living peacefully and treating others with respect is at the core of my spirituality. If we follow these principles, our community can thrive," explained a respondent. By fostering an environment of mutual respect and collaboration, the Rakhain community leverages their spiritual values to address both internal and external challenges.

Actionable Steps for Sustainability The Rakhain Indigenous community's perspectives on sustainability reflect a pragmatic and community-driven approach to environmental stewardship. This section explores their strategies for tree planting and preservation, area-based awareness, and educational and training initiatives.

Tree Planting and Preservation Tree planting is a cornerstone of the Rakhain community's sustainability efforts, addressing both ecological and practical needs. As one participant explained, "We have to tell people to plant trees, especially banyan trees, for shelter. Trees also help prevent soil erosion and maintain the natural balance." The emphasis on banyan trees underscores their multifunctionality, providing shade, stabilizing soil, and enhancing biodiversity.

The community's approach also considers the protective role of specific tree species during natural disasters. "If there is no tree, soil erosion will happen. In the monsoon, we have lightning. You can tell them to plant palm trees because they take the effects of lightning," shared a respondent. This statement highlights the importance of selecting tree species based on local environmental conditions and challenges.

The historical loss of tree cover is a recurring concern, driving reforestation efforts. "Trees like tamarisk and banyan provide protection during natural disasters. By planting these, we can create a more resilient environment," noted a participant. This vision aligns with global movements for afforestation and ecosystem restoration, demonstrating the Rakhain community's commitment to sustainable development.

Area-Based Awareness Localized awareness campaigns are critical for addressing region-specific environmental issues. The Rakhain community advocates for targeted messaging to promote sustainable practices. "Send area-based awareness messages, like planting trees at the bank of rivers in Barisal to prevent soil erosion. Also, plant trees on the dam to strengthen it," suggested a respondent. Such initiatives highlight the importance of tailoring interventions to the unique geographical and ecological contexts of each area.

The consequences of unchecked environmental degradation are particularly evident in regions like Bandarban. "In Bandarban, cutting hills and trees is leading to water scarcity. People bring water from faraway places now because local sources are destroyed," explained a participant. This example underscores the need for regulatory measures and community engagement to prevent further damage.

The loss of traditional ecological knowledge is another concern, as illustrated by changes in Kuakata. "In Kuakata, there used to be coconut and tamarisk gardens, but now there is nothing. People have destroyed these resources," lamented a respondent. Reviving such practices requires a combination of awareness, education, and policy support to reintegrate traditional knowledge into contemporary sustainability efforts.

Educational and Training Initiatives Education and skill development are central to the Rakhain community's vision for sustainability. By empowering individuals with practical skills, they aim to foster self-reliance and economic resilience. One participant shared, "We are providing training to the SSC passed boys and sewing machines to girls. This helps them learn skills that reduce dependency and empower them economically." Such initiatives bridge the gap between traditional practices and modern livelihoods, ensuring that cultural heritage informs sustainable development.

Food security is another priority, with a focus on local production. "Planting vegetables can help families become self-reliant. If people can produce their food, it can significantly reduce poverty in our community," noted a respondent. This approach aligns with global calls for sustainable agriculture, emphasizing community-driven solutions to food insecurity.

The integration of technology and modern skills is also emphasized. "Our youth need to learn skills like knitting, sewing, and computers to become more sustainable. Job opportunities are limited, so self-sufficiency is vital," explained a participant. By equipping younger generations with diverse competencies, the community prepares them to navigate a rapidly changing world while maintaining their cultural and ecological roots.

The Rakhain Indigenous community's insights into cultural identity, spirituality, and sustainability offer valuable lessons for addressing contemporary challenges. Their efforts to preserve their language and traditions highlight the importance of cultural resilience in the face of systemic aggression. Spirituality serves as both a source of strength and a guiding framework for ethical living, fostering harmony and mutual respect.

In the realm of sustainability, the community's pragmatic approaches to tree planting, area-based awareness, and education reflect a deep understanding of local contexts and challenges. By integrating traditional knowledge with modern practices, they demonstrate the potential for holistic and inclusive development.

These findings underscore the need for collaborative efforts to support Indigenous communities in preserving their cultural heritage and promoting sustainability. By amplifying their voices and incorporating their perspectives into policy and practice, we can work toward a more equitable and sustainable future.

Discussion

The findings reveal the Rakhine Indigenous community's profound understanding of environmental and spiritual interdependence, presenting critical insights into the human-created climate crisis and the role of land-based spirituality in fostering resilience. This discussion contextualizes these findings within broader South Asian scholarship, highlighting key themes such as the anthropogenic impacts on ecosystems, the integration of Indigenous ecological knowledge, and the transformative potential of spirituality in climate action.

Anthropogenic Impacts on Ecosystems The detrimental effects of human activity on the environment, as articulated by the Rakhine, align with observations across South Asia. Studies have identified deforestation, industrial pollution, and unsustainable agricultural practices as primary drivers of ecological degradation (Rahman et al., 2020). In Bangladesh, the destruction of mangroves in the Sundarbans—a crucial natural barrier against cyclones—has exacerbated vulnerability to climate disasters (Ghosh et al., 2018). Similarly, in India, unregulated mining and logging have led to biodiversity loss and increased soil erosion in Himalayan regions (Singh et al., 2017). The Rakhine's critique of human greed mirrors Vandana Shiva's (2016) argument that the commodification of nature under neoliberal policies has prioritized profit over ecological balance. Such practices, as noted by the Rakhine, disrupt the intrinsic harmony between humans and nature, resulting in rising sea levels, frequent cyclones, and biodiversity loss. This underscores the need for regional and global policies to address the systemic roots of environmental degradation while incorporating localized solutions.

Integration of Indigenous Ecological Knowledge Indigenous ecological knowledge (IEK) has been increasingly recognized as a vital resource for sustainable environmental management. The Rakhine community's practices, such as planting trees to mitigate soil erosion and conserve biodiversity, exemplify this (McGregor, 2014). Similar practices are observed among the Khasi in India, who maintain sacred groves to protect forest biodiversity (Tiwari et al., 2018). These groves function as reservoirs of genetic diversity, supporting both ecological health and cultural identity. The emphasis on reciprocal relationships with nature, as articulated by the Rakhine, resonates with the Chipko Movement in India, where communities actively protected forests from commercial logging (Guha, 1989). These practices highlight the potential of community-led conservation initiatives rooted in traditional knowledge to address ecological crises effectively. Scholars like Berkes (2018) advocate for the integration of IEK with scientific frameworks to create holistic strategies for climate adaptation. Despite its effectiveness, IEK often remains marginalized in formal policy-making. The Rakhine's insights emphasize the urgency of reversing this trend by recognizing and institutionalizing Indigenous contributions. For example, participatory approaches that co-create solutions with local communities, as implemented in Nepal's community forestry programs, have significantly enhanced forest regeneration and local livelihoods (Ojha et al., 2009).

Land-based Spirituality as Climate Action Land-based spirituality, as practiced by the Rakhine, provides a powerful framework for fostering ecological ethics and resilience. Their adherence to Panchashila—ethical precepts emphasizing non-violence and respect for nature—parallels the teachings of Mahatma Gandhi, who advocated for Ahimsa (non-violence) as a principle of harmonious coexistence with nature (Hardiman, 2003). Similarly, Buddhist monastic communities in Sri Lanka have engaged in reforestation projects, guided by spiritual values that view trees as sacred and integral to life (De Silva, 2017).

The Rakhine's reverence for trees and water aligns with the concept of "sacred ecology" observed among the Lepcha community in India, who consider rivers and mountains as sacred entities deserving protection (Arora, 2006). These spiritual frameworks inspire collective action for conservation by fostering a sense of moral responsibility and interconnectedness. Taylor (2010) argues that such spiritual ecologies can challenge the exploitative paradigms of modernity, promoting more sustainable relationships with the natural world. However, the marginalization of spiritual dimensions in mainstream climate discourse limits the potential for transformative change. The Rakhine's approach highlights the need for interdisciplinary research that bridges environmental science, anthropology, and religious studies to explore the role of spirituality in climate resilience. Such efforts can deepen our understanding of how spiritual values motivate environmental stewardship and community cohesion.

Teaching Sustainability to Future Generations The Rakhine's emphasis on intergenerational knowledge transfer addresses a critical need for building long-term resilience. Their initiatives to teach youth practical skills such as tree planting and sustainable agriculture mirror efforts in Bhutan's Gross National Happiness framework, which integrates environmental education into school curricula (Ura et al., 2012). Such programs nurture ecological consciousness from an early age, fostering a culture of sustainability.

In South Asia, educational initiatives that combine traditional knowledge with modern science have demonstrated success in enhancing community resilience. For example, the M. S. Swaminathan Research Foundation in India trains farmers in agroecology, blending Indigenous practices with scientific techniques to improve productivity and environmental health (Swaminathan, 2010). These approaches align with the Rakhine's vision of self-reliance and community empowerment through skill development. The integration of spirituality into education, as practiced by the Rakhine, adds a unique dimension to sustainability learning. By linking ecological practices with moral and spiritual values, they provide a holistic framework that addresses both practical and ethical dimensions of environmental stewardship. Future initiatives can draw inspiration from this model to design culturally relevant and impactful educational programs.

Addressing Systemic Challenges The Rakhine's experiences highlight systemic barriers to effective climate action, including inadequate government support and corruption. These challenges are pervasive across South Asia, where marginalized communities often face disproportionate impacts from climate disasters (Islam & Winkel, 2017). For instance, during Cyclone Aila, bureaucratic inefficiencies delayed relief distribution, exacerbating the suffering of affected communities in Bangladesh (Haque et al., 2019). To address these issues, inclusive governance structures are essential. Nepal's success with community forestry, which devolves management authority to local groups, demonstrates the benefits of empowering communities to take charge of their resources (Ojha et al., 2009). Similar models can be adapted to the Rakhine context, ensuring that their voices are heard in decision-making processes.

The critique of anthropocentrism in the Rakhine's philosophy calls for a reevaluation of development paradigms that prioritize economic growth over ecological balance. Shiva (2016) argues for the adoption of "Earth Democracy," which centers ecological sustainability and social justice as foundational principles. Aligning policies with such frameworks can support Indigenous communities in addressing both environmental and socio-economic challenges.

The findings from the Rakhine Indigenous community offer valuable insights into addressing the human-created climate crisis through a synthesis of ecological knowledge, spiritual ethics, and community-driven action. South Asian scholarship reinforces the significance of these perspectives, demonstrating the effectiveness of Indigenous practices and land-based spirituality in fostering resilience. By integrating these approaches into broader climate strategies, policymakers and practitioners can create more inclusive and sustainable solutions. Future research should prioritize participatory methodologies to amplify Indigenous voices and explore the intersections of spirituality, ecology, and resilience. Embracing the wisdom of communities like the Rakhine is not only an ethical imperative but also a pragmatic necessity for addressing the multifaceted challenges of climate change.

Chapter 10
Responsibilities and Moving Forward

The concluding chapter synthesizes the critical discussions of the book, addressing the intensified effects of the climate crisis and settler colonialism on Indigenous communities in Bangladesh. It critiques Western, technocratic approaches to climate adaptation, which often marginalize Indigenous perspectives and prioritize economic gains over ecological and cultural sustainability. The chapter highlights the transformative potential of Indigenous land-based practices rooted in ecological stewardship, spiritual connections, and intergenerational knowledge. These practices, demonstrated by communities like the Munda, Marma, Khyeng, and Rakhine, offer sustainable solutions for resilience and adaptation in coastal, hilly, and wetland regions. The chapter also emphasizes the importance of recognizing Indigenous land rights and self-determination as cornerstones of climate justice. It advocates for policies that integrate traditional knowledge into climate adaptation strategies, support Indigenous community-led governance, and ensure Indigenous land-based culture and practice to empower future generations. Through decolonial learning, the chapter proposes dismantling colonial legacies in climate governance and advancing inclusive, equitable frameworks that amplify Indigenous voices globally. By centering Indigenous perspectives, the book underscores the urgency of fostering a sustainable and just future, wherein Indigenous knowledge systems play a pivotal role in addressing the climate crisis.

The climate crisis continues to expose and exacerbate systemic inequities, disproportionately affecting Indigenous communities globally. All the chapters showcased how in Bangladesh Indigenous populations residing in coastal, hilly, and wetland areas are particularly vulnerable to the critical impacts of human-induced climate disasters and historical ongoing settler colonialism. This book synthesizes the critical analyses presented in previous chapters, emphasizing the shortcomings of Western perspectives on climate adaptation, the transformative potential of Indigenous land-based knowledge, and the importance of these practices in fostering Indigenous self-determination, cultural revitalization, and climate justice.

R. Datta, *Decolonizing Climate Adaptation*,
https://doi.org/10.1007/978-981-96-9140-1_10

Challenging Western Perspectives on Climate Adaptation

Bangladesh's government and NGOs' Western approaches to climate adaptation often rely on mainstream outsiders' profit-oriented frameworks, prioritizing economic growth over Indigenous environmental sustainability and cultural preservation. All the chapters discussed how these ongoing settlers' approaches frequently marginalize Indigenous voices and knowledge systems, treating them as supplementary rather than foundational. In Bangladesh, this manifests in policies that favour large-scale development projects and industrial agriculture, often at the expense of Indigenous lands and livelihoods (Datta, 2018; Roy, 2000). For example, Chap. 2 shows how mainstream forest management practices in coastal areas prioritize monoculture plantations for economic gains, neglecting the biodiversity and ecological resilience fostered by traditional Indigenous methods (Garai et al., 2022; Nath et al., 2016). Similarly, deforestation in the CHT disrupted hydrological systems and displaced Indigenous communities, illustrating how technocratic solutions exacerbate vulnerabilities rather than mitigating them (Adnan, 2004; Chowdhury, 2008). Therefore, decolonizing mainstream settlers' climate management policy and practices requires a shift from top-down, extractive governance to community-driven, culturally grounded approaches. Such a paradigm shift acknowledges the limitations of Western models and the need for land-based, relational frameworks that integrate diverse knowledge systems, including Indigenous epistemologies (Datta, 2020).

The Importance of Indigenous Land-Based Learning and Practice

Indigenous land-based practices offer sustainable and context-specific solutions to the climate crisis. Rooted in deep ecological understanding and spiritual connections to land, these practices prioritize harmony with nature, resilience, and intergenerational equity. In Bangladesh, Indigenous communities demonstrate remarkable adaptability through traditional water management systems, agroforestry, and biodiversity conservation.

Sustainable Land-Based Practices and Resilience Indigenous practices such as crop rotation, seed preservation, and rainwater harvesting enhance ecological sustainability while ensuring food security and resilience to climate shocks. For instance, the Munda Indigenous communities in coastal areas showcase how to rely on mangroves not only as natural buffers against cyclones but also as sources of livelihood and cultural identity (Garai et al., 2022; Haque et al., 2019). Similarly, Marma Indigenous communities explained their elevated housing and floating agriculture in the Chittagong Hill Tracts exemplify adaptive strategies that mitigate flooding risks and preserve traditional knowledge (Roy, 2000).

Spirituality and Stewardship As all the chapters showcase the land-based spiritual ethos of Indigenous communities plays a crucial role in their adaptive strategies. Practices such as land-based rituals, daily gratitude, and seasonal agricultural preparations reinforce a sense of collective responsibility and ecological stewardship. Khyeng Indigenous communities'land-based spirituality, for example, fosters resilience by aligning agricultural practices with natural cycles and promoting sustainable resource management (Chakma, 2016; Nath et al., 2016).

By centering these practices in climate adaptation policies, researchers and policymakers can address not only the ecological dimensions of resilience but also the cultural and spiritual values that underpin them. This holistic approach aligns with the Relational and Land-Based theoretical frameworks, which emphasize interconnectedness and cultural integrity as essential components of sustainability (Wilson, 2016).

Building Indigenous Self-Determination and Land Rights Indigenous self-determination is inextricably linked to land rights, cultural preservation, and climate justice. Secure land tenure enables Indigenous communities to exercise sovereignty over their territories, implement sustainable practices, and maintain cultural continuity. However, historical and ongoing land dispossession remains a significant barrier to these goals.

Legal Recognition and Advocacy Advocating for legal recognition of Indigenous land rights is a cornerstone of decolonial climate adaptation. All the chapters explained how ongoing land grabbing and infrastructural projects often encroach on Indigenous territories, displacing communities and undermining their adaptive capacities (Adnan, 2004; Roy, 2000). Restitution and legal protections are essential for empowering Indigenous communities to lead adaptation efforts on their own terms.

Community-Led Governance Self-determination also involves reshaping governance structures to prioritize Indigenous agencies. Community-led approaches to climate adaptation, such as participatory mapping and co-management of natural resources, ensure that Indigenous perspectives shape decision-making processes. These practices not only enhance the effectiveness of adaptation strategies but also uphold principles of equity and justice (Datta, 2018; Haque et al., 2019).

Revitalizing Traditional Culture and Practice Revitalizing traditional culture and practices is central to Indigenous resilience and climate adaptation. By preserving and celebrating Indigenous knowledge systems, communities can strengthen intergenerational ties, foster innovation, and resist cultural erosion.

Intergenerational Traditional Land-based Culture and Practice Intergenerational Indigenous traditional land-based knowledge and practices (TLBKP) learning is a cornerstone of cultural revitalization. Storytelling, mentorship, and collaborative practices ensure the transmission of traditional knowledge, fostering continuity and innovation. In the Manipuri and Rakhine communities, for example, elders play a crucial role in teaching youth about sacred connections to land, rituals, and sustainable practices, empowering them to navigate contemporary challenges while honoring their heritage (Garai et al., 2022; Nath et al., 2016).

Cultural Preservation in Policy All the studies in this research highlight the importance of TLBKP for cultural preservation in climate policies reinforcing the importance of Indigenous identity and agency. Policies that support traditional practices, such as seed banks and community-led conservation initiatives, not only enhance resilience but also affirm the cultural rights of Indigenous peoples. This approach aligns with the Indigenist and Trans-Systematic theoretical frameworks, which advocate for the integration of Indigenous knowledge into broader governance systems (Battiste, 2013; Wilson, 2016).

Advancing Climate Justice Through Decolonial Frameworks A decolonial approach to climate justice addresses the systemic inequities that disproportionately burden Indigenous communities. This involves dismantling colonial legacies, fostering equitable collaborations, and amplifying Indigenous voices in global climate governance.

Addressing Systemic Inequities and Ongoing Systemic inequities in climate governance often exclude Indigenous perspectives, perpetuating marginalization and environmental degradation. Decolonizing these systems requires inclusive policies that prioritize Indigenous sovereignty, cultural relevance, and ecological sustainability. For instance, integrating Indigenous adaptive strategies into national climate plans can enhance their effectiveness while promoting equity (Adnan, 2004; Roy, 2000).

Centering Indigenous Leadership The lessons learned from Indigenous climate adaptation in Bangladesh have broader implications for global climate governance. By centering Indigenous leadership and knowledge, policymakers can develop scalable solutions that balance ecological integrity with social justice. Community-led mangrove restoration projects in Bangladesh, for example, offer a model for coastal resilience that can inspire similar initiatives worldwide (Garai et al., 2022; Haque et al., 2019).

Therefore, this book takes political responsibility for decolonizing climate adaptation and integrating Indigenous land-based practices are critical for addressing the multifaceted challenges of the climate crisis in Bangladesh. This book highlights the need to move beyond settlers' models and embrace holistic, community-driven approaches that respect Indigenous sovereignty, cultural integrity, and ecological wisdom. By advocating for land rights, self-determination, and cultural revitalization, Indigenous communities can lead the way in building resilience and advancing climate justice. This book collectively highlights the transformative potential of Indigenous knowledge and practices. From the spiritual ethos of the Khyeng and

Manipuri communities to the adaptive innovations of the Marma and Munda, these narratives demonstrate how traditional wisdom offers actionable solutions to contemporary challenges. By integrating these perspectives into global climate governance, we can create a more inclusive, equitable, and sustainable future for all.

Decolonial Learning Reflections and Policy: Advancing Indigenous Climate Adaptation in Bangladesh

As this book explained many Indigenous communities have been facing multifaceted challenges, particularly for Indigenous communities whose resilience is deeply tied to their land, culture, and knowledge systems. Decolonial learning offers an opportunity to re-evaluate existing policies and practices through an inclusive lens that centers on Indigenous perspectives. This section explores the reflections gained from decolonial learning and provides policy recommendations to strengthen Indigenous climate adaptation in Bangladesh.

Decolonial Learning Reflections Decolonial learning begins with critically examining how colonial histories shape contemporary governance systems. In Bangladesh, colonial legacies are evident in policies that prioritize economic development over environmental sustainability and cultural preservation. Large-scale infrastructural projects, such as deforestation for monoculture plantations and dam constructions, have displaced Indigenous communities and disrupted traditional knowledge systems (Adnan, 2004; Roy, 2000). The marginalization of Indigenous voices in policymaking perpetuates inequities and undermines resilience. For instance, the exclusion of Indigenous ecological practices from national climate strategies reflects a technocratic bias that prioritizes Western scientific approaches while neglecting relational and land-based frameworks (Datta, 2018; Wilson, 2016). Decolonial learning emphasizes the need to dismantle these inequities and recognize the value of Indigenous knowledge in addressing climate challenges.

Centering Indigenous Knowledge and Land-Based Practices Indigenous land-based practices represent sustainable and adaptive solutions to climate vulnerabilities. These practices, rooted in centuries of ecological stewardship, emphasize harmony with nature and intergenerational learning. The Munda community's reliance on mangroves as natural buffers against cyclones and the Marma's elevated housing structures for flood mitigation highlight the efficacy of traditional ecological knowledge (Haque et al., 2019; Nath et al., 2016). Decolonial learning acknowledges that Indigenous spirituality and cultural practices are integral to their adaptive strategies. The rituals and seasonal agricultural preparations of the Khyeng and Manipuri communities demonstrate how spirituality fosters environmental stewardship and resilience (Chowdhury, 2008; Garai et al., 2022). Incorporating these practices into broader climate policies can enhance cultural relevance and ecological effectiveness.

Addressing Gender and Intergenerational Dimensions Decolonial learning also emphasizes the importance of gender and intergenerational dimensions in climate adaptation. Indigenous women play a central role in sustaining cultural practices and resource management, making them key agents of resilience. For example, women-led seed preservation initiatives in the Manipuri community ensure food security while promoting biodiversity (Garai et al., 2022).

Similarly, intergenerational learning fosters continuity and innovation. Elders in the Rakhine community transmit traditional knowledge to younger generations through storytelling and mentorship, empowering youth to navigate contemporary challenges while honoring cultural heritage (Nath et al., 2016). Recognizing these dynamics is essential for creating inclusive and effective adaptation strategies.

Decolonizing Climate Policy and Practices Indigenous land rights is fundamental to Indigenous self-determination and resilience. Policies must prioritize the legal recognition of Indigenous territories to prevent displacement and promote sustainable land use. In Bangladesh, restitution for historical land dispossession and protection against encroachments are critical for empowering communities to implement adaptive practices on their terms (Adnan, 2004; Roy, 2000).

Implementing TLBKP into Climate Policies National climate strategies must integrate TLBKP into ecological practices and cultural values. This requires establishing frameworks for participatory policy-making, where Indigenous leaders collaborate with policymakers to co-design adaptation initiatives. For example, incorporating Munda's mangrove conservation techniques into coastal management plans can enhance resilience while preserving cultural heritage (Haque et al., 2019).

Support for Community-Led Governance Decolonial policies should promote community-led governance models that prioritize Indigenous agencies. Participatory mapping, co-management of natural resources, and local decision-making structures ensure that adaptation strategies reflect community priorities and expertise (Battiste, 2013; Datta, 2018). These models foster equity and accountability while enhancing the effectiveness of climate interventions.

Land-based Education and Capacity Building Education initiatives must be culturally grounded to support intergenerational knowledge transmission and capacity building. Incorporating Indigenous languages, histories, and ecological practices into curricula can empower youth to engage with climate challenges from a culturally informed perspective. Collaborative learning programs that bring together Indigenous and non-Indigenous stakeholders can also foster mutual understanding and innovation (Nath et al., 2016; Wilson, 2016).

Advocacy for Inclusive Climate Justice Frameworks Climate justice frameworks must address systemic inequities by amplifying Indigenous voices in global governance. This involves advocating for Indigenous representation in international forums, ensuring that their perspectives shape global climate policies. Community-led initiatives, such as mangrove restoration projects and agroforestry systems, can serve as models for equitable and sustainable adaptation strategies worldwide (Garai et al., 2022; Haque et al., 2019).

Implementing Decolonial Policies in Practice

The success of decolonial policies depends on their implementation through collaborative and inclusive processes. Researchers, policymakers, and community leaders must work together to translate decolonial principles into actionable outcomes. This includes:

- Community-led participatory research: Led by Indigenous communities as researcher and/or co-researchers to ensure that studies reflect their priorities and knowledge systems.
- Resource allocation: Providing financial and technical support for community-led adaptation initiatives, such as reforestation and flood mitigation projects.
- Monitoring and evaluation: Developing metrics that account for cultural and ecological outcomes alongside conventional measures of success.

By embedding decolonial principles into policy and practice, stakeholders can create a more inclusive and sustainable approach to climate adaptation.

In this book, the decolonial learning reflections, underline the transformative potential of Indigenous land-based practices and knowledge systems in addressing the climate crisis. By critiquing colonial legacies, centering Indigenous perspectives, and promoting inclusive policies, stakeholders can advance climate justice while fostering cultural revitalization and ecological sustainability. The policy recommendations outlined in this chapter emphasize the importance of legal recognition, community-led governance, and culturally grounded education in strengthening Indigenous resilience. By integrating these principles into national and global frameworks, policymakers can create equitable solutions that honor Indigenous sovereignty and promote a sustainable future for all.

References

Acharya, A. (2002). Democratisation and the prospects for participatory regionalism in Southeast Asia. *Third World Quarterly, 24*(2), 375–390.

Adger, W. N. (2000). Social and ecological resilience: Are they related? *Progress in Human Geography, 24*(3), 347–364.

Adger, W. N., Barnett, J., Brown, K., Marshall, N., & O'Brien, K. (2014). Cultural dimensions of climate change impacts and adaptation. *Nature Climate Change, 3*(2), 112–117.

Adger, W. N., Hughes, T. P., Folke, C., Carpenter, S. R., & Rockström, J. (2005). Social-ecological resilience to coastal disasters. *Science, 309*(5737), 1036–1039.

Adnan, S. (2004). *Migration, land alienation and ethnic conflict: Causes of poverty in the Chittagong Hill Tracts of Bangladesh.* University Press Limited.

Agrawal, P. K. (2010). To what extent are the indigenous women of Jharkhand, India living in disadvantageous conditions: Findings from India's National Family Health Survey 1. *Asian Ethnicity, 11*(1), 61–80.

Ahmed, A. U., Haque, N., & Neelormi, S. (2019). Climate change and disaster resilience in the Chittagong Hill Tracts. *Journal of Environmental Management, 242*, 110–120.

Ahmed, A. U., Huq, S., & Nasreen, M. (2015). Climate change adaptation practices in Bangladesh: Lessons from indigenous communities. *Environmental Management, 56*(5), 105–120.

Ahmed, I., Mallick, B., & Rahman, R. (2019). Impacts of salinity intrusion on agriculture in coastal Bangladesh. *Environmental Management, 56*(4), 789–799.

Alam, M. (2017). Indigenous knowledge in climate adaptation. *Journal of Management, 6*, 1393354.

Alam, M. K., et al. (2017). Indigenous knowledge and climate adaptation in the Chittagong Hill tracts. *Journal of Environmental Management, 200*, 1–12.

Alden Wily, L. (2011). Challenging the State: Devolutionary tenure transitions for saving and expanding forests. *Human Ecology, 49*(3), 285–295.

Ali, T., Paton, D., Buergelt, P. T., Smith, J. A., Jehan, N., & Siddique, A. (2021). Integrating indigenous perspectives and community-based disaster risk reduction: A pathway for sustainable Indigenous development in Northern Pakistan. *International Journal of Disaster Risk Reduction, 59*, 102263.

Altieri, M. A. (2004). *Agroecology: The science of sustainable agriculture.* CRC Press.

Anaya, S. J. (2009). The Human Rights of Indigenous Peoples, in Light of the New Declaration, and the Challenge of Making Them Operative: Report of the Special Rapporteur on the Situation of Human Rights and Fundamental Freedoms of Indigenous People. Light of the New Declaration, and the Challenge of Making Them Operative: Report of the Special Rapporteur on the Situation of Human Rights and Fundamental Freedoms of Indigenous People (August 5, 2009).

R. Datta, *Decolonizing Climate Adaptation*,
https://doi.org/10.1007/978-981-96-9140-1

Arora, V. (2006). The forest of symbols embodied in the Tholung sacred landscape of North Sikkim, India. *Conservation and Society, 4*(1), 55–83.

Baird, R. (2008). *The impact of climate change on minorities and Indigenous peoples*. Minority Rights Group International.

Barkat, A., Zaman, S., & Raihan, S. (2010). *Political economy of land litigation in Bangladesh: A case of collusive practices and abuses of power*. University Press Limited.

Bartlett, C., Marshall, M., & Marshall, A. (2012). Two-eyed seeing and the role of indigenous knowledge in climate change adaptation. *Environmental Health Review, 65*(3), 77–82.

Battiste, M. (2002). *Indigenous knowledge and pedagogy in first nations education*. National Working Group on Education.

Battiste, M. (2013). *Decolonizing education: Nourishing the learning spirit*. Purich Publishing Limited.

Begum, H. A., Perveen, R., Chakma, E., Dewan, L., Afroze, R. S., & Tangen, D. (2019). The challenges of geographical inclusive education in rural Bangladesh. *International Journal of Inclusive Education, 23*(1), 7–22.

Berkes, F. (2008). *Sacred ecology*. Routledge.

Berkes, F. (2009). Indigenous ways of knowing and the study of environmental change. *Journal of the Royal Society of New Zealand, 39*(4), 151–156.

Berkes, F. (2018). *Sacred ecology: Traditional ecological knowledge and resource management*. Routledge.

Berkes, F., Colding, J., & Folke, C. (2000). Rediscovery of traditional ecological knowledge as adaptive management. *Ecological Applications, 10*(5), 1251–1262.

Berkes, F., & Folke, C. (1998). *Linking social and ecological systems: Management practices and social mechanisms for building resilience*. Cambridge University Press.

Bhadra, T., Sharma, A., & Singh, R. K. (2021). Climate change vulnerability in Indigenous communities of Northeast India. *Environmental Research, 195*, 110768.

Bohensky, E. L., & Maru, Y. (2011). Indigenous knowledge, science, and resilience: What have we learned from a decade of international literature on "multiple evidence bases"? *Ecology and Society, 16*(4), 16.

Bullock, R., Zurba, M., Reed, M. G., & McCarthy, D. (2023). Strategic options for more effective indigenous participation in collaborative environmental governance. *Journal of Planning Education and Research, 43*(4), 841–856.

Cajete, G. (2000). *Native science: Natural laws of interdependence*. Clear Light Publishers.

Callaghan, T. V., Kulikova, O., Rakhmanova, L., Topp-Jørgensen, E., Labba, N., Kuhmanen, L. A., & Johansson, M. (2020). Improving dialogue among researchers, local and indigenous peoples and decision-makers to address issues of climate change in the North. *Ambio, 49*, 1161–1178.

Cameron, L., Mauro, I., & Settee, K. (2021). "A return to and of the land": Indigenous knowledge and climate change initiatives across the Canadian Prairies. *Journal of Ethnobiology, 41*(3), 368–388.

Cannon, T. D., Hennah, W., van Erp, T. G., Thompson, P. M., Lonnqvist, J., Huttunen, M., … & Peltonen, L. (2003). Association of DISC1/TRAX haplotypes with schizophrenia, reduced prefrontal gray matter, and impaired short-and long-term memory. *Archives of General Psychiatry, 62*(11), 1205–1213.

Carter, J., & Hollinsworth, D. (2014). Teaching Indigenous geography in a neo-colonial world. *Journal of Geography in Higher Education, 41*(2), 182–197.

Chakma, B. (2018). Challenges in the CHT. *Asian Journal of Environment, 13*, 45–56.

Chakma, B., & Tripura, M. (2019). Land and spirituality: The cultural resilience of the Khyeng community. *Bangladesh Indigenous Studies Review, 7*(1), 12–25.

Chakma, S. (2016). Indigenous spirituality and environmental ethics: Perspectives from the Marma community. *Asian Journal of Philosophy, 18*(2), 45–60.

Chakma, S., & Chakma, R. (2021). Land-based spirituality among the Marmas: A cultural ecology perspective. *Journal of Indigenous Studies, 12*(3), 22–40.

Chapin, F. S., Zavaleta, E. S., Eviner, V. T., Naylor, R. L., Vitousek, P. M., Reynolds, H. L., & Díaz, S. (2009). Consequences of changing biodiversity. *Nature, 405*(6783), 234–242.

Chausson, A., Turner, B., Seddon, D., Chabaneix, N., Girardin, C. A., Kapos, V., & Seddon, N. (2020). Mapping the effectiveness of nature-based solutions for climate change adaptation. *Global Change Biology, 26*(11), 6134–6155.

Chongtham, N., Yumnam, S., & Laishram, J. (2018). Rituals and biodiversity conservation in Manipur, India. *Asian Journal of Environmental Studies, 3*(2), 45–56.

Chowdhury, R. (2008). Colonial impacts on indigenous peoples' adaptive capacity. *Journal of Environmental Studies, 16*(2), 123–139.

Claeys, P., & Edelman, M. (2020). The United Nations Declaration on the rights of peasants and other people working in rural areas. *The Journal of Peasant Studies, 47*(1), 1–68.

Coates, K. S. (2004). *A global history of indigenous peoples*. Palgrave Macmillan.

Corntassel, J. (2012). Re-envisioning resurgence: Indigenous pathways to decolonization and sustainable self-determination. *Decolonization, 1*, 86–101.

Coulthard, G. (2014). *Red skin, white masks: Rejecting the colonial politics of recognition*. University of Minnesota Press.

Crate, S. A., & Nuttall, M. (2009). *Anthropology and climate change: From encounters to actions*. Routledge.

Crutzen, P. J. (2006). The anthropocene: A geology of mankind. *Nature, 415*, 23–24.

Datta, R. (2015). A relational theoretical framework and meanings of land, nature, and sustainability for research with Indigenous communities. *Local Environment, 20*(1), 102–113.

Datta, R. (2018). Decolonizing both research and researcher: Critical insights from research with indigenous communities in Bangladesh. *Research Ethics, 14*(2), 1–16.

Datta, R. (2020). *Land-water management and sustainability in Bangladesh: Indigenous practices in the Chittagong Hill tracts*. Routledge.

Datta, R. (2022). Indigenous trans-systemic research approach. *Qualitative Inquiry, 28*(6), 694–702.

Datta, R. (Ed.). (2023). *Decolonization in practice: Reflective learning from cross-cultural perspectives*. Canadian Scholars' Press.

Datta, R. (2024). Relationality in Indigenous climate change education research: A learning journey from Indigenous communities in Bangladesh. *Australian Journal of Environmental Education, 40*(2), 128–142.

Datta, R., & Chapola, J. (2018). Indigenous and western environmental resource management: a learning experience with Laitu Khyeng Indigenous community, Chittagong Hill Tracts (CHT), Bangladesh. *Journal of Native American and Indigenous Studies, 4*(2), 52–80. https://doi.org/10.5749/natiindistudj.4.2.0052

Datta, R., Chapola, J., Owen, K., Hurlbert, M., & Foggin, A. (2024). Indigenous land-based practices for climate crisis adaptions. *The Explorer, 20*(6), 103042.

Datta, R., & Kairy, B. (2024). Decolonizing climate change adaptations from Indigenous perspectives: Learning reflections from Munda Indigenous communities, coastal areas in Bangladesh. *Sustainability, 16*(2), 769.

De Silva, P. (2017). Buddhist responses to environmental challenges in Sri Lanka. *Journal for the Study of Religion, Nature & Culture, 11*(2), 200–219.

Deininger, K., & Feder, G. (2001). Land institutions and land markets. *Handbook of Agricultural Economics, 1*, 287–331.

Dudgeon, R. C., & Berkes, F. (2003). Local understandings of the land: Traditional ecological knowledge and Indigenous resource management. *Human Ecology, 31*(2), 117–128.

Dumont, A., Fortin, M.-J., & Gauthier, S. (2017). Traditional ecological knowledge and resilience in indigenous communities. *Ecology and Society, 22*(4), 11–30.

Dunlap, R. E., & Brulle, R. J. (2015). *Climate change and society: Sociological perspectives*. Oxford Academic.

Dutta, P., Singh, R. K., & Sharma, T. (2019). Biodiversity loss in Northeastern India: Challenges and solutions. *Biodiversity and Conservation, 28*(6), 1681–1698.

Emmons, R. A., & McCullough, M. E. (2003). Counting blessings versus burdens: An experimental investigation of gratitude and subjective well-being in daily life. *Journal of Personality and Social Psychology, 84*(2), 377–389. https://doi.org/10.1037/0022-3514.84.2.377

Escobar, A. (1995). *Encountering development: The making and unmaking of the Third World.* Princeton University Press.

Farley, M. (2022). Making the connections: Resource extraction, prostitution, poverty, climate change, and human rights. *The International Journal of Human Rights, 26*(6), 1032–1055.

Ford, J. D., & Pearce, T. (2010). What we know, do not know, and need to know about climate change vulnerability in the Arctic: A systematic literature review. *Climatic Change, 100*(1), 121–136.

Ford, J. D., et al. (2020). A framework for assessing the vulnerability of Indigenous communities to climate change. *Global Environmental Change, 64*, 102114.

Gadgil, M., Berkes, F., & Folke, C. (1993). Indigenous knowledge for biodiversity conservation. *Ambio, 22*(2-3), 151–156.

Gaillard, J. C., Maceda, E. A., Stasiak, E., Le Berre, I., & Espaldon, M. V. O. (2007). Sustainable livelihoods and people's vulnerability in the face of coastal hazards. *Journal of Coastal Conservation, 13*(2), 119–129.

Garai, J., Rasul, G., & Ullah, M. (2022). Exploring Indigenous land management practices for sustainable agriculture in South Asia. *Asian Agricultural Research, 14*(3), 45–58.

Ghosh, S., Biswas, J., & Sharma, D. (2018). Mangrove afforestation: A sustainable solution to cyclone impact in the Sundarbans. *Journal of Coastal Research, 85*, 1164–1171.

Guha, R. (1989). *The unquiet woods: Ecological change and peasant resistance in the Himalaya.* University of California Press.

Gupta, A., & Sharma, S. (2020). Climate justice: Indigenous perspectives and global imperatives. *Climate Policy, 20*(8), 1008–1021.

Hall, A., & Patrinos, H. A. (2006). *Indigenous peoples, poverty, and development*. Cambridge University Press.

Hamilton, C. (2013). *Earthmasters: The dawn of the age of climate engineering*. Yale University Press.

Hansen, J., Sato, M., & Ruedy, R. (2013). *Global temperature update through 2012*. NASA Goddard Institute for Space Studies.

Haque, U., Hashizume, M., Kolivras, K. N., Overgaard, H. J., Das, B., & Yamamoto, T. (2019). Reduced death rates from cyclones in Bangladesh: What more needs to be done? *Bulletin of the World Health Organization, 90*, 150–156.

Hardiman, D. (2003). Gandhi's environmentalism: Moral and political issues. *Economic and Political Weekly, 38*(4), 305–314.

Hauser, M. (2018). Cultural resilience and Indigenous knowledge in climate adaptation. *Sustainability Science, 13*(5), 1207–1218.

Henry, E., & Pene, H. (2001). Kaupapa Māori: Locating indigenous ontology, epistemology, and methodology in the academy. *Organization, 8*(2), 234–242.

Hill, J. (2019). Indigenous knowledge for climate change adaptation and resilience in rural communities of Ankole Sub-region, Uganda. *African Journal of Climate Change and Resource Sustainability, 4*(1), 239–251.

Hossain, M. A., & Chakma, S. (2018). Environmental ethics in Marma spiritual practices. *International Journal of Cultural Studies, 10*(4), 105–120.

Huq, S. (2015). Policy and practice in Bangladesh. *Environmental Policy, 25*, 10.

Huq, S., Reid, H., & Murray, L. A. (2013). Community-based adaptation: A vital approach to climate change. *Climate and Development, 5*(3), 167–174.

Huq, S., et al. (2015). Climate change and Bangladesh: Policy and practice. *Environmental Policy and Governance, 25*(3), 1–10.

IPCC. (2021). *Climate change 2021: The physical science basis. Contribution of Working Group I to the Sixth Assessment Report of the Intergovernmental Panel on Climate Change*. IPCC.

Islam, M. R., Sallu, S., Hubacek, K., & Paavola, J. (2019). Vulnerability of Indigenous communities in the Sundarbans, Bangladesh. *Global Environmental Change, 29*, 199–207.

Islam, M. R., & Winkel, J. (2017). Climate change and social vulnerability in Bangladesh. *Climate Change, 8*(1), 17–33.

Islam, M. S., Rahman, M. H., & Haque, A. (2014). Vulnerabilities of indigenous communities in the Chittagong Hill tracts. *Journal of Disasters and Resilience, 6*(2), 84–100.

Islam, M. S., & Walkerden, G. (2014). How do links between households and NGOs promote disaster resilience and recovery? *Natural Hazards, 69*(3), 1587–1603.

Jackson, S., & Moggridge, B. (2019). Indigenous water management. *Australasian Journal of Environmental Management, 26*(3), 193–196.

Jasanoff, S. (2018). *The fifth branch: Science Advisers as policymakers*. Harvard University Press.

Justrabo, C. (2023). *Restoration of traditional marine food systems as climate adaptation strategy for coastal First Nations* (Doctoral dissertation, University of British Columbia).

Khan, A. (2025). Biopolitics as a weapon of land-grabbing: Exploring land contention between a Rubber plantation company and three Indigenous communities in Bangladesh. *International Journal on Minority and Group Rights, 1*(aop), 1–23.

Khan, A. R., Mahi, M. H., Mojumdar, M. U., Chakraborty, N. R., Kabir, M. A., & Boruah, A. N. (2021, April). Transformer-Based Model for Identifying Ethnic Languages in Bangladesh. In 2025 12th International Conference on Computing for Sustainable Global Development (INDIACom) (pp. 1–6). IEEE.

Khumbongmayum, A. D., Khan, M. L., & Tripathi, R. S. (2005). Sacred groves of Manipur, northeast India: Biodiversity value, status, and strategies for their conservation. *Biodiversity and Conservation, 14*(7), 1541–1582.

Kimmerer, R. W. (2013). *Braiding sweetgrass: Indigenous wisdom, scientific knowledge, and the teachings of plants*. Milkweed Editions.

King, U. (2013). *Spirituality and environmental ethics: Nature, spirit, and the human future*. Cambridge University Press.

Kipp, A., Cunsolo, A., Gillis, D., Sawatzky, A., & Harper, S. L. (2019). The need for community-led, integrated and innovative monitoring programmes when responding to the health impacts of climate change. *International Journal of Circumpolar Health, 78*(2), 1517581.

Kothari, A., Pathak, N., & Vania, F. (2013). *Sharing power: Learning by doing in co-management of natural resources throughout the world*. Earthscan.

Kothari, A., et al. (2019). *Pluriverse: A post-development dictionary*. Tulika Books.

Koubi, V. (2019). *Sustainable development impacts of climate change and natural disaster*. Background Paper Prepared for Sustainable Development Outlook.

Kumar, A., Kumar, S., Komal, Ramchiary, N., & Singh, P. (2021). Role of traditional ethnobotanical knowledge and indigenous communities in achieving sustainable development goals. *Sustainability, 13*(6), 3062.

LaDuke, W. (2016). *Recovering indigenous land sovereignty*. Yale University.

Lama, D. (2020). *Our only home: A climate appeal to the world*. Hanover Square Press.

Leal Filho, W., Lütz, J. M., Totin, E., Ayal, D., & Mendy, E. (2018). Obstacles to implementing indigenous knowledge in climate change adaptation in Africa. *Journal of Environmental Management, 373*, 123678.

Leichenko, R., & O'Brien, K. (2008). *Environmental change and globalization: Double exposures*. Oxford University Press.

Maldonado, J., Wang, I. F. C., Eningowuk, F., Iaukea, L., Lascurain, A., Lazrus, H., & Thomas, B. (2021). Addressing the challenges of climate-driven community-led resettlement and site expansion: Knowledge sharing, storytelling, healing, and collaborative coalition building. *Journal of Environmental Studies and Sciences, 11*(3), 294–304.

Mallick, B., & Rahman, A. (2017). Cyclone shelters and their role in disaster risk reduction in Bangladesh. *Environmental Hazards, 16*(2), 165–183.

Marma Community Interview (2023) will be Learned from the Marma Community conversation.

Mansoor, S., Farooq, I., Kachroo, M. M., Mahmoud, A. E. D., Fawzy, M., Popescu, S. M., & Ahmad, P. (2022). Elevation in wildfire frequencies with respect to the climate. *Journal of Environmental Management, 301*, 113769.

Mason, L. R., & Rigg, J. (Eds.). (2019). *People and climate change: Vulnerability, adaptation, and social justice*. Oxford University Press.

McGregor, D. (2014). Traditional ecological knowledge and sustainability. *Canadian Journal of Native Education, 37*(1), 11–36.

Mercer, J., Kelman, I., Taranis, L., & Suchet-Pearson, S. (2010). Framework for integrating indigenous and scientific knowledge for disaster risk reduction. *Disasters, 34*(1), 214–239.

Mehran, A., AghaKouchak, A., Nakhjiri, N., Stewardson, M. J., Peel, M. C., Phillips, T. J., & Ravalico, J. K. (2017). Compounding impacts of human-induced water stress and climate change on water availability. *Scientific Reports, 7*(1), 6282.

Muthukrishnan, R., & Datta, R. (2023). *Indigenous practice and community-led climate change solutions: The relevance of traditional cosmic knowledge systems*. Taylor & Francis.

Nakashima, D., Prott, L., & Bridgewater, P. (2000). Tapping into the world's wisdom. *UNESCO Sources, 125*, 12–15.

Nakashima, D., & Roué, M. (2002). Indigenous knowledge and environmental ethics: A review of international policy discussions. *Ethics in Science and Environmental Politics, 9*(1), 19–32.

Nakashima, D. J., Galloway McLean, K., Thulstrup, H. D., Ramos Castillo, A., & Rubis, J. T. (2012). *Weathering uncertainty: Traditional knowledge for climate change assessment and adaptation*. UNESCO.

NASA. (2021). *The causes of climate change*. Retrieved from https://climate.nasa.gov/causes/

Nath, B., Chowdhury, R., & Datta, R. (2016). Traditional ecological knowledge in disaster resilience: Lessons from Bangladesh. *Indigenous Knowledge Systems, 5*(1), 89–105.

Nelson, A. R., & Wilson, M. (2018). Indigenous governance and climate change mitigation in the arctic.

Ojha, H., Persha, L., & Chhatre, A. (2009). Community forestry in Nepal: A policy innovation for local livelihoods and food security. *International Forestry Review, 11*(2), 197–207.

Posey, D. A. (1999). Safeguarding traditional resource rights of indigenous peoples. *Ethnoecology: Situated knowledge/located lives* (pp. 217–229). The University of Arizona Press.

Pretty, J., & Smith, D. (2004). Social capital in biodiversity conservation and management. *Conservation Biology, 18*(3), 631–638.

Quinn, P., Williamson, B., & Gibbs, L. (2022). Indigenous-informed disaster recovery: Addressing collective trauma using a healing framework. *Progress in Disaster Science, 16*, 100257.

Rahman, A., Alam, M., & Chakma, S. (2012). Disaster resilience and cultural heritage in indigenous communities. *Journal of Environmental Studies, 8*(1), 35–50.

Rahman, A., & Mallick, B. (2022). Indigenous strategies to climate resilience in Bangladesh. *Regional Environmental Change, 22*(3), 56–78.

Rahman, M. H. (2020). Indigenous knowledge in disaster risk reduction: Lessons from the Marmas. *Journal of Humanitarian Assistance, 12*(3), 44–60.

Rahman, M. H., & Alam, K. (2016). Forest dependent indigenous communities' perception and adaptation to climate change through local knowledge in the protected area—A Bangladesh case study. *Climate, 4*(1), 12.

Rahman, M. S., Haque, A., & Jahan, S. (2020). Resilience of coastal communities to climate change: Evidence from Bangladesh. *Environmental Management, 65*(3), 497–509.

Rasmussen, M. B., Räikkönen, J., & Vadstrup, M. (2021). Spiritual ecology: Bridging spirituality and environmental ethics. *Environmental Values, 30*(2), 211–230.

Reed, M. S., Evely, A. C., Cundill, G., Fazey, I., Glass, J., Laing, A., & Stringer, L. C. (2018). Building resilience through adaptive co-management. *Environmental Management, 54*(3), 529–540.

Robbins, P. (2012). *Political ecology: A critical introduction*. Wiley-Blackwell.

Roy, A. (2018). Sacred groves. *Journal of Forestry, 13*, 170–181.

Roy, A. (2020). *Water management*. Water Policy.

Roy, A., et al. (2018). Sacred groves and biodiversity conservation in the chittagong hill tracts. *Journal of Forestry Research, 29*(4), 789–798.
Roy, A., et al. (2020). Community-based water management in the Chittagong Hill tracts. *Water Policy, 22*(6), 1–15.
Roy, J., & Chan, E. (2019). Strengthening climate resilience through Indigenous practices. *Sustainability, 11*(5), 4323.
Roy, R. C. K. (2000). *Land rights of the indigenous peoples of the Chittagong Hill tracts, Bangladesh*. IWGIA.
Schilling-Vacaflor, A. (2017). 'If the company belongs to you, how can you be against it?' Limiting participation and taming dissent in neo-extractivist Bolivia. *The Journal of Peasant Studies, 44*(3), 658–676.
Schipper, E. L. F., Dubash, N. K., & Mulugetta, Y. (2020). Climate change research and the search for solutions: Rethinking interdisciplinarity. *Climatic Change, 168*(3), 18.
Sponsel, L. E. (2024). Religion and sustainability: Social movements and the politics of the environment. *Journal of Ecological Anthropology, 17*(1), 33–36.
Seddon, N., Chausson, A., Berry, P., Girardin, C. A., Smith, A., & Turner, B. (2020). Understanding the value and limits of nature-based solutions to climate change and other global challenges. *Philosophical Transactions of the Royal Society B, 375*(1794), 20190120.
Shaffril, H. A. M., Ahmad, N., Samsuddin, S. F., Samah, A. A., & Hamdan, M. E. (2020). Systematic literature review on adaptation towards climate change impacts among indigenous people in the Asia Pacific regions. *Journal of Cleaner Production, 258*, 120595.
Sharma, R., Singh, S., & Devi, M. (2017). Agricultural rituals and sustainability in Manipuri culture. *Cultural Ecology Journal, 12*(2), 34–56.
Shaw, D. (2017). Bridging the gap: Two-eyed seeing and climate adaptation. *Fish and Fisheries, 22*, 243–261.
Shiva, V. (2016). *Earth democracy: Justice, sustainability, and peace*. North Atlantic Books.
Simpson, L. B. (2014). Land as pedagogy: Nishnaabeg intelligence and relearning education. *Decolonization, 3*, 1–25.
Singh, R. K., Singh, A., & Yumnam, K. (2020). Sustainable farming practices in Indigenous communities of Northeast India. *Agroecology and Sustainable Food Systems, 44*(6), 763–776.
Singh, S. P., Rawat, G. S., & Verma, D. (2017). Biodiversity conservation in the Indian Himalayas: Trends, challenges, and opportunities. *Biodiversity, 18*(4), 151–164.
Smith, H. A., & Sharp, K. (2012). Indigenous climate knowledge in adaptation. *Climatic Change, 120*(1), 241–257.
Smith, J., Brown, K., & Adger, W. N. (2019). Indigenous resilience to climate change. *Annual Review of Environment and Resources, 44*(1), 409–429.
Smith, L. T. (1999). *Decolonizing methodologies: Research and indigenous peoples*. Zed Books.
Smith, L. T. (2012). *Decolonizing methodologies: Research and indigenous peoples*. Zed Books.
Sopher, D. E. (1968). Pilgrim circulation in Gujarat. *Geographical Review*, 392-425.
Steffen, W., Persson, Å., Deutsch, L., Zalasiewicz, J., Williams, M., Richardson, K., & Svedin, U. (2011). The anthropocene: From global change to planetary stewardship. *Ambio, 40*, 739–761.
Steffen, W., et al. (2015). The trajectory of the Anthropocene: The great acceleration. *Anthropocene Review, 2*(1), 81–98.
Sultana, F. (2021). Climate change and Indigenous women: A feminist analysis. *Climate Policy, 21*(6), 867–877.
Swaminathan, M. S. (2010). *From green to evergreen revolution: Indian agriculture at the crossroads*. Academic Foundation.
Tanchangya, P. (2020). Seasonal rituals and ecological knowledge in indigenous communities. *Asian Ethnology, 79*(2), 1–22.
Taylor, B. (2010). *Dark green religion: Nature spirituality and the planetary future*. University of California Press.

Tilman, D., Cassman, K. G., Matson, P. A., Naylor, R., & Polasky, S. (2002). Agricultural sustainability and intensive production practices. *Nature, 418*(6898), 671–677.

Tiwari, B. K., Barik, S. K., & Tripathi, R. S. (2018). Sacred forests of Meghalaya: Biological and cultural significance. *Journal of Forestry Research, 22*(3), 491–500.

Trott, C. D., Lam, S., Roncker, J., Gray, E. S., Courtney, R. H., & Even, T. L. (2023). Justice in climate change education: A systematic review. *Environmental Education Research, 29*, 1535–1572.

Tu'itahi, S., Watson, H., Egan, R., Parkes, M. W., & Hancock, T. (2021). Waiora: The importance of Indigenous worldviews and spirituality to inspire and inform planetary health promotion in the anthropocene. *Global Health Promotion, 28*(4), 73–82.

Tuck, E., & Yang, K. W. (2012). Decolonization is not a metaphor. *Decolonization, 1*, 1–40.

UNICEF. (2020). *The role of youth in climate change education*. UNICEF.

Ura, K., Alkire, S., & Zangmo, T. (2012). GNH and happiness: Understanding Bhutan's national index of wellbeing. *Social Indicators Research, 110*(3), 783–801.

Verlie, B. (2021). *Learning to live with climate change: From anxiety to transformation*. Routledge.

Vogel, B., Yumagulova, L., McBean, G., & Charles Norris, K. A. (2022). Indigenous-led nature-based solutions for the climate crisis: Insights from Canada. *Sustainability, 14*(11), 6725.

Walker, B., Holling, C. S., Carpenter, S. R., & Kinzig, A. (2004). Resilience, adaptability, and transformability in social-ecological systems. *Ecology and Society, 9*(2), 5.

Wals, A. E., Brody, M., Dillon, J., & Stevenson, R. B. (2014). Convergence between science and environmental education. *Science, 344*(6184), 583–584.

Watson, J., Singh, V., & Davis, A. (2021). Climate justice in practice: Indigenous perspectives. *Geoforum, 120*, 73–83.

Watts, V., Hooks, G., & McLaughlin, N. (2018). A troubling presence: Indigeneity in English-language Canadian sociology. *Canadian Review of Sociology/Revue Canadienne de Sociologie, 57*(1), 7–33.

Whyte, K. (2013). *On the role of indigenous knowledge in climate change adaptation*.

Whyte, K. P. (2017). Indigenous climate change studies: Indigenizing futures, decolonizing the Anthropocene. *English Language Notes, 55*(1-2), 153–162.

Whyte, K. P. (2018). Indigenous climate justice. *Human Ecology, 46*(3), 431–436.

Whyte, K. P. (2020). Against crisis epistemology. *Globalizations, 17*(4), 635–649.

Wilson, E. O. (2016). *Half-earth: Our planet's fight for life*. Liveright Publishing Corporation.

Wilson, S. (2008). *Research is ceremony. Indigenous research methods*. Fernwood.

Wolfe, P. (2006). Settler colonialism and the elimination of the native. *Journal of Genocide Research, 8*(4), 387–409.

Zavala, M. (2013). What do we mean by decolonizing research strategies? Lessons from decolonizing, Indigenous research projects in New Zealand and Latin America. *Decolonization: Indigeneity, Education and Society, 2*(1), 55–71.

Zurba, M., & Papadopoulos, A. (2023). Indigenous participation and the incorporation of indigenous knowledge and perspectives in global environmental governance forums: a systematic review. *Environmental Management, 72*(1), 84–99.

Index

R. Datta, *Decolonizing Climate Adaptation*,
https://doi.org/10.1007/978-981-96-9140-1

MIX
Papier aus verantwortungsvollen Quellen
Paper from responsible sources
FSC® C105338

If you have any concerns about our products,
you can contact us on
ProductSafety@springernature.com

In case Publisher is established outside the EU,
the EU authorized representative is:
Springer Nature Customer Service Center GmbH
Europaplatz 3, 69115 Heidelberg, Germany

Printed by Libri Plureos GmbH
in Hamburg, Germany